A Simulation Tool to Assess Contaminant Warning System Sensor Performance Characteristics

Awwa
Research
Foundation

About the Awwa Research Foundation

The Awwa Research Foundation (AwwaRF) is a member-supported, international, nonprofit organization that sponsors research to enable water utilities, public health agencies, and other professionals to provide safe and affordable drinking water to consumers.

The Foundation's mission is to advance the science of water to improve the quality of life. To achieve this mission, the Foundation sponsors studies on all aspects of drinking water, including supply and resources, treatment, monitoring and analysis, distribution, management, and health effects. Funding for research is provided primarily by subscription payments from approximately 1,000 utilities, consulting firms, and manufacturers in North America and abroad. Additional funding comes from collaborative partnerships with other national and international organizations, allowing for resources to be leveraged, expertise to be shared, and broad-based knowledge to be developed and disseminated. Government funding serves as a third source of research dollars.

From its headquarters in Denver, Colorado, the Foundation's staff directs and supports the efforts of more than 800 volunteers who serve on the board of trustees and various committees. These volunteers represent many facets of the water industry, and contribute their expertise to select and monitor research studies that benefit the entire drinking water community.

The results of research are disseminated through a number of channels, including reports, the Web site, conferences, and periodicals.

For subscribers, the Foundation serves as a cooperative program in which water suppliers unite to pool their resources. By applying Foundation research findings, these water suppliers can save substantial costs and stay on the leading edge of drinking water science and technology. Since its inception, AwwaRF has supplied the water community with more than $300 million in applied research.

More information about the Foundation and how to become a subscriber is available on the Web at **www.awwarf.org.**

A Simulation Tool to Assess Contaminant Warning System Sensor Performance Characteristics

Prepared by:
Wayne Einfeld and **Sean A. McKenna**
Sandia National Laboratories
P.O. Box 5800, Albuquerque, NM 87185-0734
and
Mark P. Wilson
GRAM and Associates
8500 Menaul Blvd. NE,, Suite B-335, Albuquerque, NM 87112

Jointly sponsored by:
Awwa Research Foundation
6666 West Quincy Avenue, Denver, CO 80235-3098
and
Sandia National Laboratories
P.O. Box 5800, Mail Stop 0755, Albuquerque, New Mexico, 87185-0755

Published by:

Awwa Research Foundation

Distributed by:

American Water Works Association

IWA Publishing

DISCLAIMER

This study was funded by the Awwa Research Foundation (AwwaRF) and Sandia National Laboratories (Sandia). AwwaRF and Sandia assume no responsibility for the content of the research reported in this publication or for the opinions or statements of fact expressed in this report. The mention of trade names for commercial products does not represent or imply the approval or endorsement of either AwwaRF or Sandia. This report is presented solely for informational purposes.

ISBN 978-1-84339-263-7

CONTENTS

TABLES

FIGURES

FOREWORD

The Awwa Research Foundation is a nonprofit corporation that is dedicated to the implementation of a research effort to help utilities respond to regulatory requirements and traditional high-priority concerns of the industry. The research agenda is developed through a process of consultation with subscribers and drinking water professionals. Under the umbrella of a Strategic Research Plan, the Research Advisory Council prioritizes the suggested projects based upon current and future needs, applicability, and past work; the recommendations are forwarded to the Board of Trustees for final selection. The foundation also sponsors research projects through the unsolicited proposal process; the Collaborative Research, Research Applications, and Tailored Collaboration programs; and various joint research efforts with organizations such as the U.S. Environmental Protection Agency, the U.S. Bureau of Reclamation, and the Association of California Water Agencies.

This publication is a result of one of these sponsored studies, and it is hoped that its findings will be applied in communities throughout the world. The following report serves not only as a means of communicating the results of the water industry's centralized research program but also as a tool to enlist the further support of the nonmember utilities and individuals.

Projects are managed closely from their inception to the final report by the foundation's staff and large cadre of volunteers who willingly contribute their time and expertise. The foundation serves a planning and management function and awards contracts to other institutions such as water utilities, universities, and engineering firms. The funding for this research effort comes primarily from the Subscription Program, through which water utilities subscribe to the research program and make an annual payment proportionate to the volume of water they deliver and consultants and manufacturers subscribe based on their annual billings. The program offers a cost-effective and fair method for funding research in the public interest.

A broad spectrum of water supply issues is addressed by the foundation's research agenda: resources, treatment and operations, distribution and storage, water quality and analysis, toxicology, economics, and management. The ultimate purpose of the coordinated effort is to assist water suppliers to provide the highest possible quality of water economically and reliably. The true benefits are realized when the results are implemented at the utility level. The foundation's trustees are pleased to offer this publication as contribution toward that end.

David E. Rager
Chair, Board of Trustees
Awwa Research Foundation

Robert C. Renner, P.E.
Executive Director
Awwa Research Foundation

ACKNOWLEDGEMENTS

The authors would like to thank David Hart for his development efforts associated with the change detection algorithms used in this investigation. We would also like to thank Victoria Cruz for running the many simulations through the change detection module, collating the results, and her assistance with troubleshooting and operations. The authors also recognize and extend their appreciation to Jennifer Warner, AwwaRF Project Manager, as well as the following AwwaRF Project Advisory Committee members for their expertise and constructive guidance and reviews throughout the course of this project.

Leonard Casson, PhD, Associate Professor, University of Pittsburg, School of Engineering

Rich Donner, Senior Engineer, Seattle Public Utilities

Eric Koglin, Project Manager, National Homeland Security Research Center, U.S. Environmental Protection Agency

Yves Mikol, PhD, Section Chief, Early Warning Remote Monitoring, Bureau of Water Supply, New York City Department of Environmental Protection

Sandia is a multiprogram laboratory operated by Sandia Corporation, a Lockheed Martin Company, for the United States Department of Energy under contract DE-AC04-94AL85000.

EXECUTIVE SUMMARY

INTRODUCTION

Water utilities are faced with the ongoing need to provide safe, contaminant-free water to their customers. The most common cause of chemical or biological contamination within water distribution systems is associated with crossed connections with waste water flows or other untreated and potentially contaminated water. Other contamination sources can include compromised source water that penetrates treatment barriers or intentional contamination attacks from disaffected groups or terrorist organizations. Residual disinfectant levels within a water distribution system can in some circumstances ameliorate the effects from such attacks; but in others the residual disinfectant levels can be totally depleted without complete inactivation of the pathogen or toxic chemical that has been injected, thereby posing a public health risk. In nearly all contamination event scenarios, early detection and response can help reduce public health risk and impact. To that end, contaminant warning systems (CWS) offer advantages over more traditional means of contaminant detection.

This research project is specifically focused on the testing of various candidate CWS sensor configurations in combination with event detection software through simulation methods. A CWS design evaluation tool has been developed that enables an assessment of overall system performance under various contamination scenarios using simulation techniques. Using such design tools, a variety of contamination scenarios can be simulated using hydraulic flow models and utility-specific pipe layouts. The location and types of sensors can then be explored to assess cost, operations and maintenance burden as well as risk reduction benefits likely to be derived from installation and use of the overall CWS system.

Project objectives were to develop a simulation tool to assess overall CWS performance under various contamination event scenarios and to incorporate specific sensor performance characteristics in the tool to more realistically assess how sensor operational parameters can influence overall system performance.

The diagram shown in Figure ES-1 conceptually illustrates how the sensor simulator functions in concert with a change detection module as the two principal components of a CWS. The input data stream is a contaminant pulse with an added water quality baseline. The water quality baseline is the background signal from water quality measurements made in the absence of a contaminant peak and includes the day-to-day changes in water quality normally encountered. The sensor simulator is configured to mimic the performance characteristics of a specific total organic carbon or residual chlorine sensor with respect to key operational parameters such as sensor noise, drift, and response time. The raw data signal is processed through the sensor simulator and the extent to which the raw signal is influenced by the sensor operational parameters is observed on the output side of the sensor simulator.

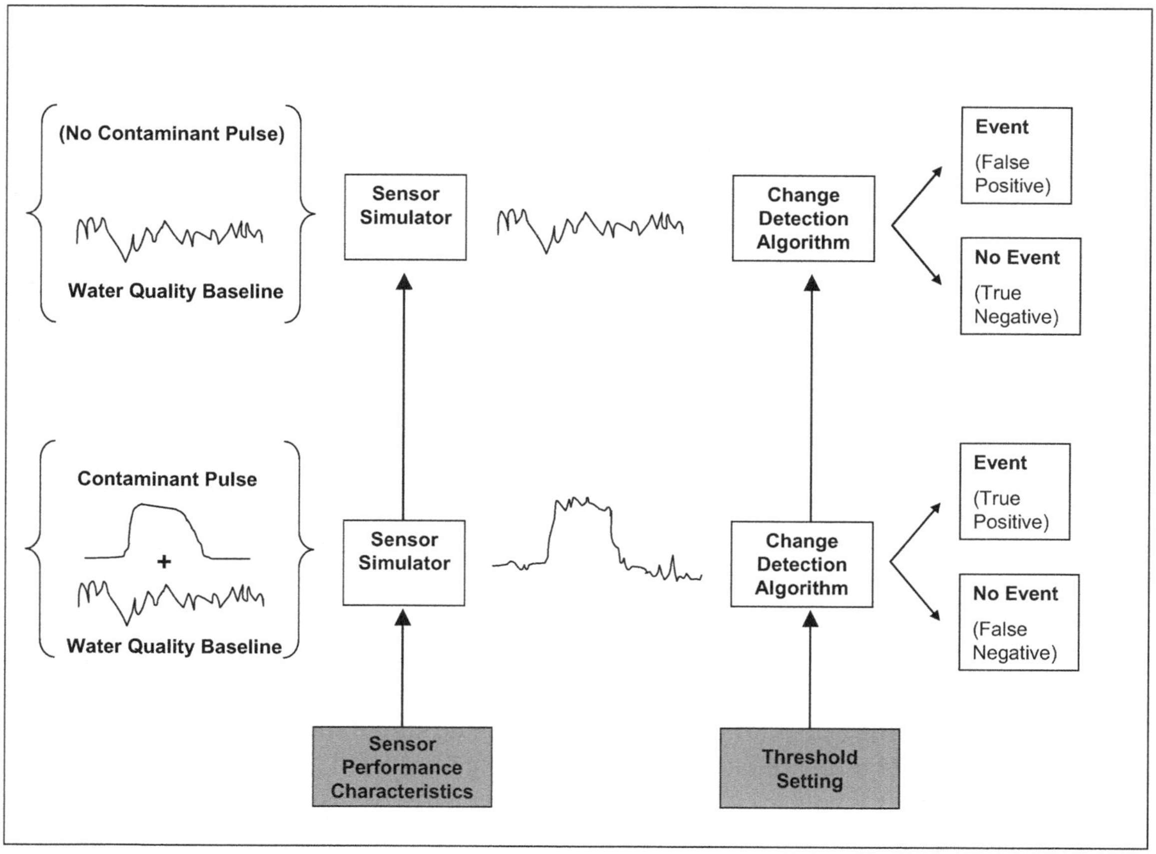

Figure ES-1. A schematic diagram showing flow of raw data through a sensor simulator and change detection algorithm.

METHODS

This study concentrated on two sensor types: residual chlorine and total organic carbon as they have been shown to be the most useful in providing surrogate measurements for a broad spectrum of contamination agents. Vendor-provided sensor performance specifications were used to configure the sensor simulator for the contamination scenarios considered in this study. A summary of the step-by-step simulation process is summarized below:

- Conduct EPANET runs for short and long pulse contamination events using the pipe layouts for a real water distribution system. EPANET is a computer model used to predict water flow and contaminant concentrations in water distribution systems.
- Obtain contaminant pulse shape at selected locations downstream from the contaminant injection point using the EPANET model.
- Linearly scale the downstream pulse shape to a pre-determined dose level for aldicarb and sodium fluoroacetate, the two contaminants selected for study.
- Remap observed concentration levels to either total organic carbon or residual chlorine sensor response using published sensor response functions.

- Superimpose a series of scaled contaminant pulses onto a water quality baseline data obtained from utility measurements. Add the contaminant pulse and baseline data together to produce a composite signal.
- Process this combined signal through the sensor simulator with sensor performance characteristics set for the sensor of interest. Sensor sampling rate, noise level, drift and other factors will alter the signal.
- Iteratively process the output signal from the sensor simulator through the event change detection module. Vary the threshold setting of the change detection module with each iteration and compile the detection statistics (e.g., true positive, false positive, true negative, false negative) for each threshold setting.
- Compile the data from the iterative runs into a format suitable for analysis using a Receiver Operating Characteristics (ROC) approach. Further analysis of ROC data serves as an indicator of overall sensor/algorithm performance.
- Repeat the simulations for other concentration levels, sensor types, contaminant pulse lengths, and baseline conditions.

The ROC curve, illustrated in Figure ES-2, is a graphical representation of the frequency of true positive detects (TP) versus the frequency of false positives (FP) over the threshold or sensitivity setting range of the particular sensor/event detection module being evaluated. For each simulation run and at a specific threshold setting, a contaminant peak is either detected (true positive) or missed (false negative). Additionally, the sensor/event detection module may detect a peak when in fact there is not one present (false positive) or correctly report a no-detect (true negative). At each event detection module threshold setting, occurrences of all of these four test results are compiled using a sensor data stream that has contaminant peaks interspersed with sections of no peaks.

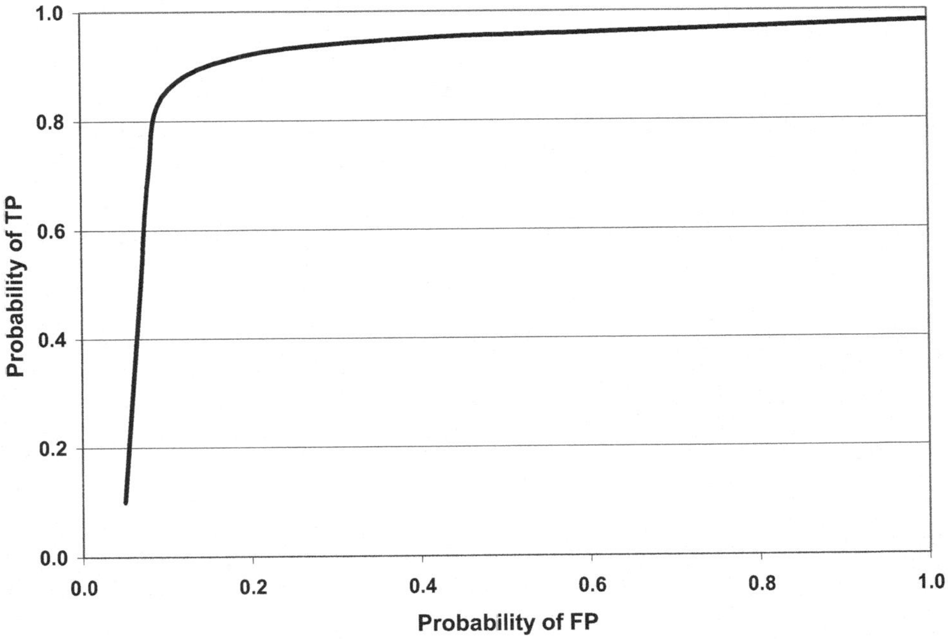

Figure ES-2. Illustration of a Receiver Operating Characteristics (ROC) curve.

RESULTS

Summaries of residual chlorine and total organic carbon sensor performance in combination with two event detection modules are shown in Tables ES-1 and ES-2. The FLP (forward linear predictor) technique involves the calculation and analysis of a moving average of the sensor data whereas the threshold method simple looks for an exceedence of a preset alarm point in the sensor data. The values tabulated are areas under the ROC curve, a statistic that gives an indication of overall system performance. Values near 0.5 reveal very poor sensor/change detection module performance with values near 1.0 indicating very good performance. Study results are summarized for a contamination scenario with aldicarb, a carbamate pesticide of moderate toxicity. Simulation results shown here involve a 30-minute contaminant injection at three dose levels with inclusion of a background sensor baseline taken from water utility measurements. The highest dose is just below the no-health-effect threshold for aldicarb. The two lower dose levels serve as surrogate cases for more toxic classes of contaminants that might be encountered in a contamination event.

Table ES-1
Performance Measures of Two Residual Chlorine Sensors
In Response to Various Contamination Events

Residual Chlorine Sensor	In-Pipe Aldicarb Concentration (mg/L)	Change Detection Method	
		FLP	Threshold
HF Scientific	High – 31	0.96	0.99
Model: CLX	Mid – 1.2	0.67	0.70
	Low – 0.05	0.52	0.56
Endress Hauser	High – 31	0.98	0.99
Model: CCS141	Mid – 1.2	0.83	0.67
	Low – 0.05	0.56	0.53

Table ES-2
Performance Measures of Four Total Organic Carbon Sensors
In Response to Various Contamination Events

TOC Sensor	In-pipe Aldicarb Concentration (mg/L)	Change Detection Method	
		FLP	Threshold
Hach	High – 31	0.88	1.00
Model: Astro TOC	Mid – 1.2	0.68	0.80
	Low – 0.05	0.55	<0.50
S:Scan	High – 31	0.97	0.98
Model: Spectro:Lyser	Mid – 1.2	0.56	0.63
	Low – 0.05	0.52	0.50
Shimadzu	High – 31	0.91	1.00
Model: TOCN-4110	Mid – 1.2	0.70	0.69
	Low – 0.05	0.52	<0.50
Endress Hauser	High – 31	0.97	0.99
Model: CSS70	Mid – 1.2	0.74	0.74
	Low – 0.05	0.52	0.51

As expected, sensor/detection module performance degraded with decreasing contaminant concentration levels. At the highest dose level, performance of all systems was relatively good. At the lowest dose level examined, all sensor/change detection combinations provided essentially no value (ROC curve areas at or below 0.5) as reliable detection devices in a CWS. In this case the probability of a false positive is essentially equal to the probability of a true positive. Differences in sensor system performance were observed and the most influential operational parameter of the sensor was the sampling interval. Those sensors with a higher frequency sampling rate performed better than those with lower sampling rates.

The event detection algorithms performed equally well for the dose levels investigated. While the more advanced statistical change detection algorithms performed on a par with the simple threshold method in this simple case, they are expected to outperform the threshold approach when used with a CWS design that includes multiple sensor types and locations. The limited scope of this study did not include the evaluation of more complex CWS designs involving multiple sensor types and locations within a water distribution system.

This study illustrates the potential advantages of simulation tools that can be used during the design of a contaminant warning system for a water utility. Through the use of such tools, inter-comparisons of candidate sensor performance can be made under a variety of realistic contamination scenarios that will enable system designers to fully optimize the warning system design prior to component acquisition, installation and deployment.

CHAPTER 1
INTRODUCTION

RESEARCH CONTEXT

Water utilities are faced with the ongoing need to provide safe, contaminant-free water to their customers. Water contamination events, although relatively rare in the U.S., have the potential to cause widespread illness and death to user populations. The most common cause of chemical or biological contamination within water distribution systems is associated with crossed connections with waste water flows or other untreated and potentially contaminated water. Other contamination sources can include source water that penetrates treatment barriers or intentional contamination attacks from disaffected groups or terrorist organizations. It is commonly understood that distribution systems are particularly vulnerable to contamination attacks since contaminant injection can occur on the downstream side of normal treatment barriers. Residual disinfectant levels within the distribution system can in some circumstances ameliorate such attacks but in others the residual disinfectant levels can be totally depleted without complete inactivation of the pathogen or toxic chemical that has been injected, thus posing a public health risk. In nearly all contamination event scenarios, early detection and response can help reduce public health risk and impact. To that end, contaminant warning systems (CWS) offer advantages over more traditional means of contaminant detection.

Utility alert and response options for distribution system contamination events fall into several categories. With respect to alerts, utilities can utilize customer complaints of altered water taste or reported observations of potentially suspect activities. However, taste and odor complaints while possibly indicative of gross contamination, may not be reliable for detection of high-toxicity chemicals or biotoxins or pathogens at infective concentration levels that are sufficiently dilute such that water taste is not altered. Utilities can also rely on continuous physical surveillance activities like video images or intrusion detection sensors at key water assets such as tanks and pump houses; however, as noted previously, contaminant injection can occur at virtually any location throughout the distribution system that is not under continuous surveillance. Other indices of exposure such as syndromic surveillance may also be available; however, these may result in significant lags in time from the actual contamination event to the alert; and particularly so if the disease incubation time in the exposed population is lengthy.

THE CONTAMINANT WARNING SYSTEM CONCEPT

Another more direct approach that can be taken by water utilities is the installation of a contaminant warning system (CWS) to provide continuous monitoring of water quality through the use of a network of sensors that are designed to yield a continuous estimate of water quality throughout a distribution system.

The concept of a CWS incorporates a number of key elements as shown in Figure 1.1. The starting point of a CWS is typically a network of sensors that directly sample treated water within the distribution system. In light of the wide spectrum of contaminants that might be encountered in a contamination event, the approach most likely to be successful is the use of indicator species or parameters that can provide an alert to a significant change in the “baseline” water quality state that may warrant further investigation. Ideally, such an alert would be

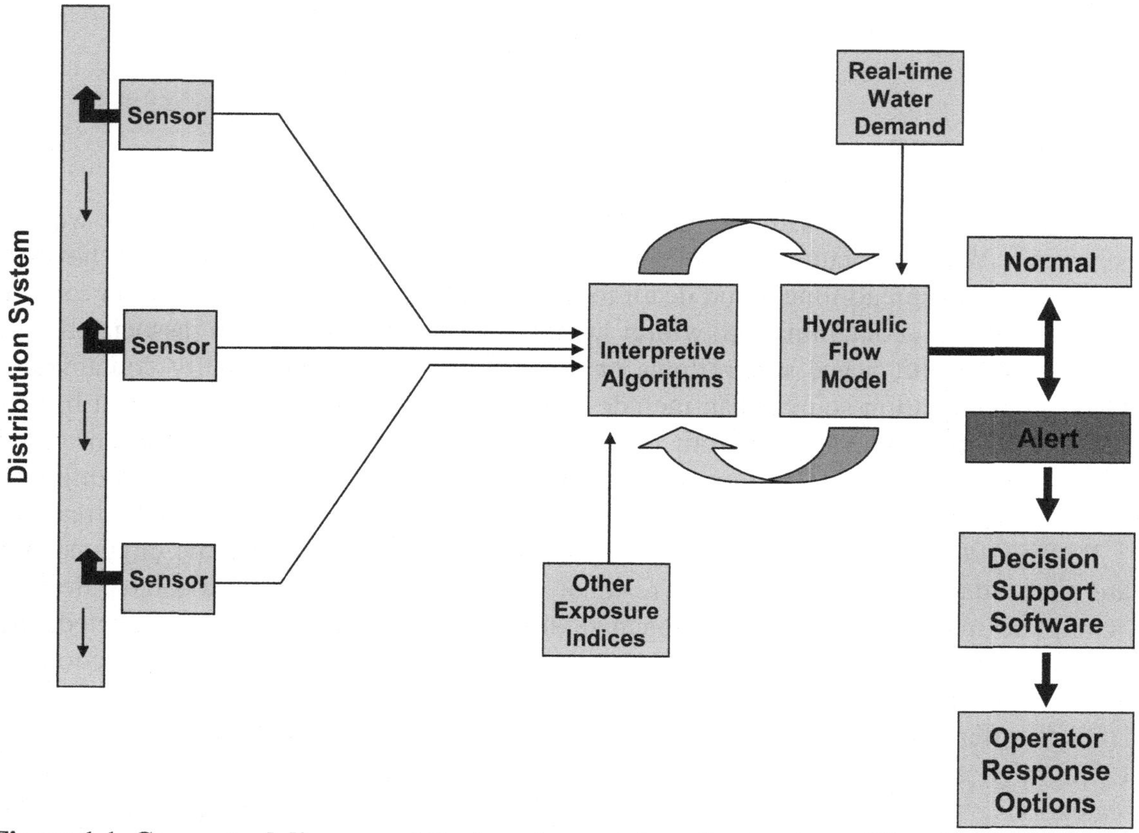

Figure 1.1 Conceptual diagram of the key elements in a Contaminant Warning System.

followed with a second-tier analysis that would provide additional detail with respect to the contaminating species; however, the current state of on-site second-tier technology is generally insufficient to provide the desired level of detail in a short time frame. A number of reviews have been published that categorize commercially available sensors for application in a CWS (EPA 2005a, Foran and Brosnan 2000, Conio et al. 2002, CBRTA 2004, States et al. 2004). A network of sensors can be utilized to provide additional confidence in the declaration of an "alert." Significant baseline deviations encountered at more than one sensor location provide additional confidence and can help to minimize the occurrence of false positives.

Surrogate sensors suitable for use in a CWS at this point in time are limited to those historically used for routine water quality assessment. Measured water quality parameters that could be linked to a contamination event include pH, turbidity, oxidative-reductive potential or ORP, residual chlorine and total organic carbon (TOC). Furthermore, software routines are typically included in a CWS that enable baseline "state change" detection with increased reliability over simple threshold exceedences. These algorithms apply statistical methods involving one or more sensors at multiple locations to increase the degree of confidence in the designation of an "alert" condition. Additional discussion of specific statistical routines used in this particular research effort occurs later in this report.

Ideally, a CWS will also include a hydraulic model of the distribution system being measured that runs in parallel and in concert with the sensor data detection algorithms. The hydraulic model allows prediction of contaminant transport throughout the distribution system as a function of time and can be used to corroborate sensor detects that would be expected to occur at a later point in time downstream from an initial detection point. The most high-fidelity hydraulic models would ideally also include a dynamic means of estimating water demand at any particular time of the day since water demand directly influences water flow volume and velocity throughout the system. In a functioning CWS, the hydraulic model can be further used in a predictive mode to indicate the likely spread of contaminant both spatially and temporally. Networked sensor data when combined with system flow data can be used in data inversion routines to gain information on the likely contaminant injection zone within the system. The degree to which the hydraulic model is calibrated to the current state of the system will impact how much information it will be able to provide during a contaminant event.

Other measures of potential exposure may also be included in a CWS. These data might include a running tally of customer complaints tabulated by distribution zone. Other sources of data may include those that fall into the category of "syndromic surveillance," such as quasi-real-time records of hospital admissions or pharmaceutical product sales at local drug stores. These sources of data, when combined with more direct measures of water quality, can provide additional insight into the likelihood of a contamination event.

Finally, a decision support software module may be a component in future systems to help the system operator make time-critical decisions following a contamination event. The decision support software is essentially an automated decision tree that takes real-time data into account, giving the system operator response options to minimize the spread of contamination to system users. Typical response options can include: physical isolation of particular storage tanks or zones within the system, flushing the system in the direction of low consumption zones or diverting dilution flow in from other storage zones that are known to be uncontaminated.

CONTAMINANT WARNING SYSTEM DEVELOPMENT CHALLENGES

At present, contaminant warning systems are in a state of infancy within the water utility industry. The US EPA is presently conducting a variety of pilot studies of contaminant warning systems in various stages of development and implementation through its Water Security Initiative (EPA 2007c). Results from these pilot studies are expected to provide additional insight into the overall feasibility of such systems for reducing public exposure risk in contamination events and the development of software tools for improved detection reliability. Various challenges faced in the further development of CWS for use within the water industry are further summarized below.

System Cost

CWS hardware can be costly both in terms of capital assets and ongoing operations and maintenance requirements. Sensor costs may range from thousands to tens of thousands of dollars with further costs associated with data acquisition systems and network linking to a central control facility. System acquisition costs are typically surpassed by operation and maintenance costs and many utilities have limited technical staff for routine operation of these systems.

Detection Reliability

At present CWS systems typically rely on surrogate measures of water quality to detect a notable change that may warrant further investigation. The potential for false positive alarms can be quite high and high false positive rates are undesirable from an operations standpoint as they tie up personnel resources and can undermine utilities' credibility with their customer base. The development of robust event detection software that can serve to dramatically reduce false positive rates is an important step in improving CWS performance reliability.

Detection Specificity

In light of the broad contaminant threat spectrum that must be considered, the current "indicator" sensor approach can quite possibly yield a reliable alert notice to the system operator; however, it is expected that additional information or a so-called "second tier" analysis will be required prior to the execution of any substantive utility response to the contamination event. For example, it is unlikely that a water system operator would issue a "do not drink" notice based on the results of a positive indication from a CWS. To that end, the development of CWS systems must be accompanied by the development of fast, field-based analytical methods that can provide a timely "second-tier" analysis in order to more fully characterize the contamination event. At present, a few utilities have access to field portable instrumentation systems (e.g., field portable gas chromatographs/mass spectrometers for chemical toxin identification, portable polymerase chain reaction systems for bio-agent identification, immunochemistry screening kits for protein toxins or bio-agent screening) to conduct these follow-on analyses. Most utilities rely upon the collection of samples followed by expedited laboratory analyses which may result in considerable time delay before quantitative data on the nature of the potential contamination are available.

Contaminant Threat Spectrum

The potential contaminant threat spectrum is quite broad and can potentially range from toxic industrial chemicals of moderate toxicity to highly toxic biotoxins or pathogens with very low infective doses. This broad threat spectrum poses a challenge for any CWS. Broad coverage with a minimum set of sensors is desired; however, given the current state of the art in sensor technology, this is goal is not yet realized.

Dual-use Opportunities

To be most effective, contaminant warning systems must provide value to the utility operator beyond the narrow goal of detection of malevolent contamination events. Information that systems can provide regarding the management of water quality, water age, system-wide disinfectant residual, cross-over contamination and other parameters in addition to security events will significantly advance the likelihood of adoption of such systems.

Model-ready Network Design Information

Expectations are that any future CWS systems will require the use of hydraulic flow models that continuously run in the background, providing a diagnostic and interpretative role during contamination events or triggers from a deployed network of sensors within the distribution system. The inclusion of such a model within a CWS necessarily implies that an adequate understanding of the layout of the utility's pipe network is required for overall CWS functionality. For large utilities, this typically does not pose a burden and the requisite investments are made to get adequate pipe design data for use in hydraulic flow models. This may not be the case for intermediate and smaller utilities; and getting to an adequate level of understanding of pipe network layout can pose significant financial burdens. Even when the pipe network is well understood, demand patterns on a fine temporal scale remain uncertain and the real-time calibration of the network model poses additional challenges.

System Design

At present, tools for the design of contaminant warning systems are limited. Utilities with an interest in equipping distribution systems with sensors will necessarily be constrained in terms of the numbers and types of sensors that they can deploy. To that end, design tools that enable informed choices as to the types, numbers and optimal locations of sensor deployment will be helpful to the industry. This research effort was focused on the development and investigation of a modeling and sensor simulation tool that can aid the designer of a CWS in the evaluation of sensor selection options in light of anticipated performance and risk reduction in an actual distribution system under a variety of threat scenarios.

STUDY OBJECTIVES

This research project is specifically focused on the final point in the foregoing discussion. A key objective in this study was the development of a CWS design evaluation tool that enables an assessment of sensor-detection algorithm system performance under various contamination scenarios through simulation methods. At present the simulation tool is limited to testing at a single location within a network; however further development of this tool could incorporate simultaneous testing at multiple locations. Presently, very few design tools enable evaluation of CWS configurations by system designers prior to actual installation. A compelling argument can be made for the need to trial various CWS sensor options and design configurations within a simulation environment. A simulation capability is appealing in that it allows a variety of CWS design alternatives to be considered in terms of overall risk reduction prior to actual system acquisition and installation. Using such design tools, a variety of contamination scenarios can be simulated using hydraulic flow models. The location and types of sensors can then be explored to assess cost, operations and maintenance burden as well as risk reduction benefits likely to be derived from installation and use of the overall system.

A second key objective in this effort is the inclusion of specific sensor performance data in the system evaluation. Until recently, much contaminant warning system development research presumed a "perfect" sensor—that is, attributes of the sensor such as detection level, electronic noise, baseline drift, sampling rate and others have been ignored or treated with generalized performance assumptions. Failure to take sensor performance attributes into account introduces considerable error into simulation results. For example, a sensor that collects a water sample every 10 minutes may miss a contaminant pulse that moves past the sampling point when the sensor is not actively sampling. The focus of this project is to incorporate actual sensor performance parameters into the simulation environment for a more accurate representation of how that particular sensor will influence the overall performance of the CWS.

CANDIDATE SENSORS FOR SIMULATION TOOL DEVELOPMENT

As noted previously, CWS systems typically rely upon water quality sensors to provide a data stream that can be used to detect significant changes in water quality that may indicate contamination. A variety of continuous or quasi-continuous sensor types are available from a number of vendors and include: pH, oxidative reductive potential, turbidity, temperature, dissolved oxygen, total chlorine, conductivity or salinity, residual free chlorine, UV absorption and total organic carbon. Limited research has been conducted on the ability of these sensor types to indicate the presence of intentional contaminants and current results show that residual chlorine and total organic carbon measurements offer the best potential among all the sensor types for a broad-class sentinel sensor approach Earlier AwwaRF-sponsored research has cataloged the variety of water quality sensors available for on-line monitoring applications (Haught and Fabris 2002, Conio et al. 2002).

For this study, we have selected representative sensor performance characteristics that reflect the various measurement methods used in both TOC and residual chlorine sensors as shown in Table 1.1.

Table 1.1
Residual Chlorine and Total Organic Carbon
Sensor Measurement Principles

Sensor Type	Measurement Principal	Investigated in this Study
Residual Chlorine	**Colorimetric** – reaction of colorimetric reagent (N,N Diethylene-p-phenylenediamine [DPD] reaction with free chlorine followed by spectrophotometric measurement.	YES
	Iodometric – reaction of potassium iodide with free chlorine to produce iodine – measures both free and complexed chlorine. Suitable for total chlorine measurements only.	NO
	Polarographic/Amperometric – electrochemical measurement of free chlorine. With combined with a free chlorine-permeable membrane, provides a measure of residual chlorine.	YES
Total Organic Carbon	**Low Temperature Oxidation** – CO2 gas removal followed by chemical and UV oxidation of sample. Detection as CO2 by non-dispersive infra-red spectroscopy	YES
	High Temperature Oxidation – CO2 gas removal followed by high temperature oxidation of the sample to CO2. Detection by NDIR or electrochemical methods.	NO
	UV Absorption – Single or double beam absorption measurements in the visible and/or UV range. Algorithm used to convert absorption data to equivalent total organic carbon.	YES

Time and funding constraints in this project did not allow a complete evaluation of all sensor types; however the tools generated in this project move in the direction of enabling the testing of additional sensor types. We have selected only those residual chlorine and TOC sensor measurement techniques that are commonly used in water distribution monitoring. Other sensor measurement methods are included in the table for comparison; however, they are typically used in other specialized applications (e.g., the high-temperature oxidation TOC method is commonly used for ultra-low TOC measurements in pharmaceutical process water).

KEY SENSOR PERFORMANCE PARAMETERS

Following is a tabulation of sensor operational parameters that are influential in the overall performance of a sensor as formally described by others (Lynggaard-Jensen 2002). Manufacturers typically do not provide published performance metrics for all of the following parameters; however, key sensor attributes such as noise, sample timing and drift are usually provided.

Random noise

The random variability of a sensor output signal over the measurement range of a sensor with a fixed input. Noise may be expressed as a standard deviation in units of raw voltage or appropriate engineering units (e.g., ± millivolts or ± mg/L free Cl). This is a general noise specification and is usually a combination of all contributing noise sources. The noise pattern is assumed to be uncorrelated in time and drawn from a Gaussian distribution.

Repeatability

The degree of variation encountered when repeated measurements of a known sample or calibrant solution are carried out. Repeatability can be expressed in standard deviation units (raw voltage or engineering units) or relative standard deviation units (i.e. the percent ratio of the standard deviation to the average value of the measurement series).

Accuracy

The "trueness" of a measurement relative to a reference standard. Accuracy is usually expressed as the percent ratio of the measured value to a reference value.

Drift

The non-random change in sensor output over time with a fixed input. Drift may be expressed in units of percent of full scale output per unit time or engineering units per unit time. (e.g., ±1% of full scale reading per hour or ± 0.1 mg/L free Cl per hour)

Sampling Interval

The frequency of sample collection and/or a measurement cycle. All sensors that are not continuous monitors have a sampling interval that can range from samples per second to samples per hour. Continuous sensors have sampling rates of 2 per second or less while sampling sensors rates range from 1.5 to 8 minutes per sample.

Response time

The time interval required for a sensor to exhibit a response to a step change in the input signal. Response time is often expressed as the time needed to reach a percent of the maximum sensor response to an input signal (e.g., 3 minutes for 90% of maximum response to a step change in the input signal).

Duty cycle

The fraction of time that the sensor is available for on-line measurement. Duty cycle may be influenced by a finite sample collection interval, other sample processing steps, downtime for calibration and maintenance, or other sensor design factors that influence its "on-line" time. The duty cycle may be expressed as a fraction of continuous time that the sensor is functional. (e.g., a 95% duty cycle implies the sensor is unavailable for measurements 5% of the time.)

Limit of Detection

The limit of detection is typically defined as the input concentration required to elicit a sensor output response that is statistically distinguishable from the baseline noise of the sensor. A signal that is greater than three times the standard deviation of the random baseline noise is most often cited as the detection criteria. The baseline noise is further defined as the standard deviation of a signal that is 5 percent of the working range of the sensor. This detection level definition is associated with a single sensor and is different from a system-wide detection level which may be defined using other statistical criteria and derived from a networked system with multiple sensor types.

Limit of Quantification

The limit of quantification is typically defined as a signal that is ten times the standard deviation or noise of the system. Signals at or above the quantification limit can be assigned a concentration level with a relatively high degree of confidence whereas a signal level that is between the detection level and the limit of quantification can be difficult to quantify with any measure of confidence.

Sensitivity

The ratio of the change in concentration of the parameter measured to the change in sensor output. This relationship may or may not be linear over the working range of the instrument. Sensor sensitivity is determined through calibration with known concentrations of the parameter being measured. (e.g., the change in mg/L free Cl per change in millivolt signal output). This parameter may also be referred to as lowest detectable change.

Specificity

Specificity is a characteristic of the sensor whereby it may respond to contaminants other than those contaminants for which it is intended. For example, a sensor designed to respond to free chlorine may also respond to nitrate ions. Specificity is typically determined through performance testing with cross contaminants and is often expressed as a ratio of the concentration a particular cross contaminant to the measured response in expected sensor output units. (e.g. , 1 mg/L of nitrate yields a sensor response equivalent to 0.1 mg/L free chlorine)

Other Factors

Other sensor performance factors that may influence sensor data quality in more subjective ways include: sensor ruggedness, resilience to varying environmental conditions like temperature, vibration and shock; sensor memory effect, the inability of the sensor to return to a previous baseline following exposure to a contaminant; and maintenance burden, the extent of sensor down time for maintenance, repair or calibration. These and other factors are not dealt with explicitly in this project.

VENDOR-PROVIDED SENSOR SPECIFICATIONS

Sensor vendors provide varying degrees of information regarding sensor performance parameters in their specification sheets and these data can be used as inputs in a sensor performance simulation. Typical performance specifications include measurement range, accuracy and/or precision, response time, sample cycle time, detection level, resolution and sensor drift. There is often no common understanding or technical definitions provided for the performance parameters of interest and testing details for how these parameters were measured may not be given. Third party verification of vendor's published sensor performance specification is desirable. However, time and cost constraints associated with performance testing often limit the availability of third-party data. There are a number of third-party testing programs that provide third-party assessments of instrument performance. The US EPA Environmental Technology Verification (ETV) Program has tested a variety of water quality sensor systems under a set of rigorous test protocols (EPA 2006). A related testing program is the EPA's Technology Test and Evaluation Program (TTEP) which is specifically focused on sensors and other technologies directly applicable to homeland security applications (EPA 2007b). The Instrumentation Testing Association is an industry-sponsored organization that conducts performance testing on water quality sensors used in the drinking water and waste water treatment industries (ITA 2006). Similar testing organizations exist in Europe, for example, the International Instrumentation Evaluation Agreement Group is focused on broad classes of sensors widely used in industry (Gluck, Palmer and Ross 2007). In this study we have limited ourselves to the use of vendor-published sensor performance specifications with the understanding that in most cases vendor-provided performance data were not verified through third-party testing.

CHAPTER 2
EXPERIMENTAL METHODS

STUDY APPROACH

As previously noted, study objectives were to develop a simulation tool to assess overall CWS performance under various contamination event scenarios and to incorporate specific sensor performance characteristics in the tool to more realistically assess how sensor operational parameters can influence overall system performance. The diagram shown in Figure 2.1 conceptually illustrates how the sensor simulator functions in concert with a change detection module as the two principal components of a CWS. The input data stream is a contaminant pulse with an added water quality baseline. The water quality baseline is the background signal from water quality measurements made in the absence of a contaminant peak and this baseline incorporates the day-to-day changes in water quality that are normally encountered.

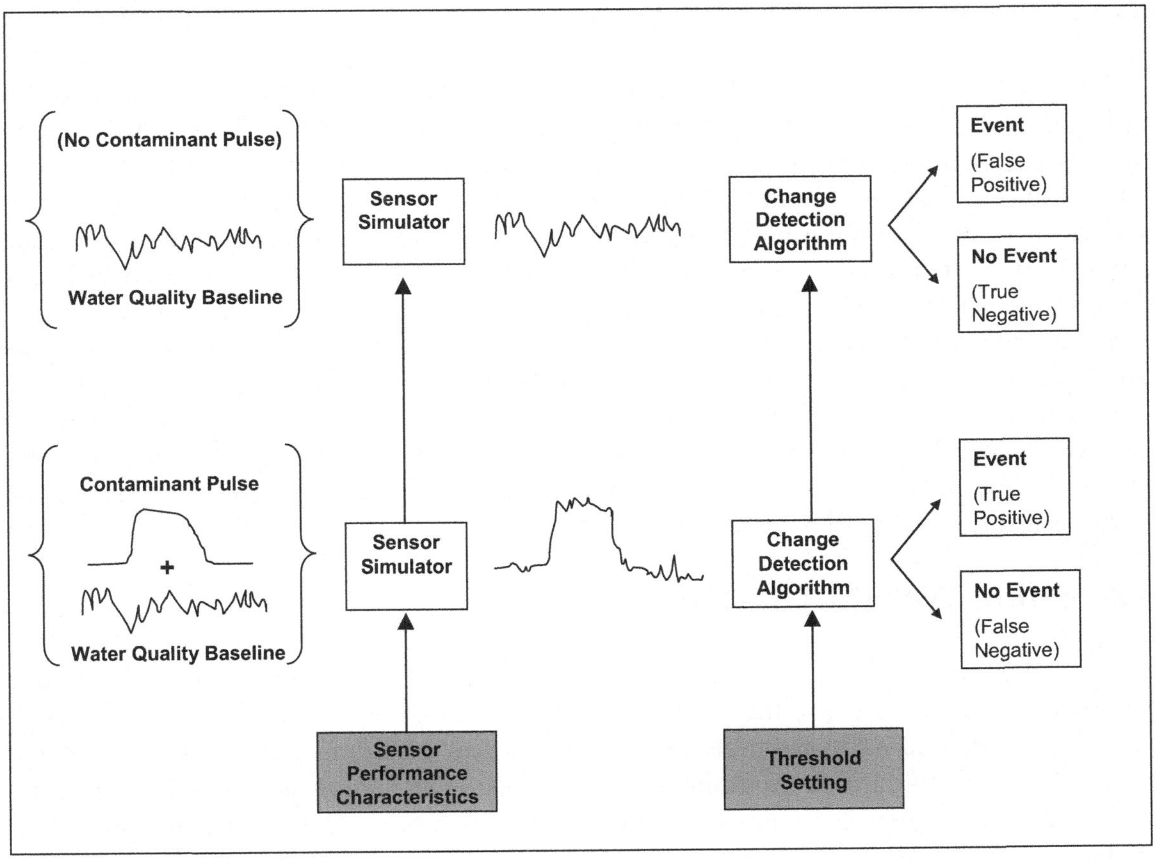

Figure 2.1 Schematic diagram of the sensor simulation process. The four possible outcomes are shown on the right, depending on whether a contaminant peak is present or not. In each case, the raw signal on the left will be altered by the sensor. This altered signal is then passed on to the change detection software in order to determine the presence or absence of a contaminant signal.

The sensor simulator is configured to mimic the performance characteristics of a specific TOC or residual chlorine sensor with respect to key operational parameters such as sensor noise, drift, and response time. The raw data signal is processed through the sensor simulator and the extent to which the raw signal is influenced by the sensor operational parameters is observed on the output side of the sensor simulator.

A second module incorporates change detection algorithms that work in conjunction with sensor data to identify contamination events of interest. The sensor output signal is processed through an event detection algorithm and the extent to which a particular sensor type influences the ability to detect the event is determined. The results at the output side of the change detection algorithm can be categorized as a true positive, false positive, false negative or true negative and are further defined in Table 2.1. Assessing the frequency of each of these four possible outcomes under specific contamination scenarios and sensor configurations enables comparison of the overall performance of various sensor options within this simulation environment.

SENSOR TYPES AND PARAMETERS INVESTIGATED

This study concentrates on two sensor types: residual chlorine and total organic carbon. Testing by the EPA Water Test and Evaluation Center in Cincinatti has shown that these two water quality parameters offer the best potential as surrogate indicators of a contamination event (Hall et al. 2007). Tables 2.2 and 2.3 include descriptions of the sensors evaluated in this research effort for residual chlorine and total organic carbon, respectively. The tables include a description of the operating principle of each sensor along with performance specifications published by the sensor manufacturer. These performance specifications are used to configure the sensor simulator for the contamination scenarios considered in this study. In most cases performance attributes are taken from the performance specification sheets published by the sensor manufacturer. Ideally, it is desirable to verify these performance parameters through independent testing; however, to do so entails significant financial and personnel resources beyond the scope of this project. For this effort we used the manufacturer-provided specifications as provided with no further verification testing.

SENSOR SIMULATOR ALGORITHMS

The sensor simulator was built using a variety of algorithms and routines available from a commercially available math routine software package (Mathworks, Natick, MA). High-level language and interactive software tools allows the user to perform computationally intensive tasks faster than with traditional programming languages. The sensor simulator also includes a graphical user interface that facilitates simplified entry of sensor performance characteristics and test conditions by the user. A variety of algorithms were incorporated into the simulator to handle key sensor attributes as further discussed below:

Table 2.1
Response Outcomes for Sensor/Algorithm Signal Analysis

Response	Condition	Sensor/Algorithm Result
True Positive (TP)	Contamination peak present	Detected
False Positive (FP)	No contamination peak present	Detected
False Negative (FN)	Contamination peak present	Not Detected
True Negative (TN)	No contamination peak present	Not Detected

Table 2.2
Residual Chlorine Sensors Selected for Study

Sensor Manufacturer	Sensor Model No.	Operating Principle	Performance Specifications
Endress Hauser	CCS140/141	Electrochemical with polymeric membrane	• Measurement Range: Model 140 (0-20 ppm or mg/L), Model 141 (0-5 ppm) • Sampling Rate/Cycle Time: 2 samples/second • Response Time: 30 seconds • Response %: 90 • Noise Value: 0.5% of measured value • Cross Contaminant %: Not provided • Drift Amount: <0.5% of sensor range • Drift Period/Unit Time: 1 month • Resolution: 0.01 mg/l
HF Scientific, Inc.	CLX	Colorimetric	• Range: 0-10 mg/l • Sampling Rate/Cycle Time: 1 sample per 90 seconds • Response Time: N/A • Response %: N/A • Noise Value: +/- 5% • Cross Contaminant %: Not provided • Drift Amount: 0 • Drift Period/Unit Time: 0 • Resolution: 0.01 mg/l

Sensor Sampling Rate and Cycle Time

The sampling rate can be varied from continuous to periodic at specified sampling intervals. This feature can accommodate sensors that may take a grab sample of water at a periodic sampling rate. Short pulses of contaminants may be missed by these types of sensors if the contaminant pulse width is less than the sensor sampling cycle. The ability to specify a continuous in-pipe sensor with no sample extraction and analysis cycle is also included. An illustration of the effect that sensor sampling rate can have on the output signal from a sensor is shown in Figure 2.2.

Sensor Response Time

An algorithm has also been implemented that addresses sensor response time for continuous sensors. Sensors do not change their output state immediately when an input change occurs. The sensor output will typically show some time lag, referred to as the response time. For example, the response time for a sensor to indicate 90 percent of the final reading of a step change in residual chlorine concentration from zero to 10 mg/L may be 3 minutes. The simulator can be configured for sensor response time by specifying the time it takes for the sensor to reach some percentage of total response in the presence of a steady-state input signal. Sensors with slower response characteristics may either underreport or miss a narrow contaminant pulse.

Table 2.3
Total Organic Carbon Sensors Selected for Study

Sensor Manufacturer	Sensor Model No.	Operating Principle	Performance Specifications
Shimadzu	TOCN-4110	Combustion/infrared analysis	• Range: 0-5 ppm, 0-1,000 ppm • Sampling Rate/Cycle Time: 1 sample per 4 minutes • Response Time: N/A • Response %: N/A • Noise Value: +/- 2% Full Scale • Noise Sigma: 1 • Cross Contaminant %: Not provided • Drift Amount: +/- 2% Full Scale • Drift Period/Unit Time: 1 week • Resolution: .01 ppm
Endress Hauser	CSS70	UV Absorption	• Range: 0.4-60 ppm, 20-900 ppm • Sampling Rate/Cycle Time: 1 sample per 40 seconds • Response Time: 60 Seconds • Response %: 90 • Noise Value: 2.5% of upper range • Noise Sigma: 1 • Cross Contaminant %: Not provided • Drift Amount: -2% Full Scale • Drift Period/Unit Time: 1 month • Resolution: .01 ppm
HACH	Astro TOC	UV-assisted oxidation/Infrared analysis	• Range: 0 – 5 up to 20,000 mg/l • Sampling Rate/Cycle Time: 1 sample per 8 minutes • Response Time: 8 minutes • Response %: 90 • Noise Value: +/- 4% of full scale • Noise Sigma: 1 • Cross Contaminant %: Not provided • Drift Amount: 2% full scale • Drift Period/Unit Time: 1 month • Resolution: .001 mg/L
S:Scan	Spectro:Lyser TOC	UV Absorption	• Range: 0 – 20 mg/L • Sampling Rate/Cycle Time: 1 sample per 30 seconds • Response Time: N/A • Response %: N/A • Noise Value: +/- 0.3 mg/L • Noise Sigma: 1 • Cross Contaminant %: Not provided • Drift Amount: 0.2 mg/L • Drift Period/Unit Time: 1 month • Resolution: 0.001 mg/L

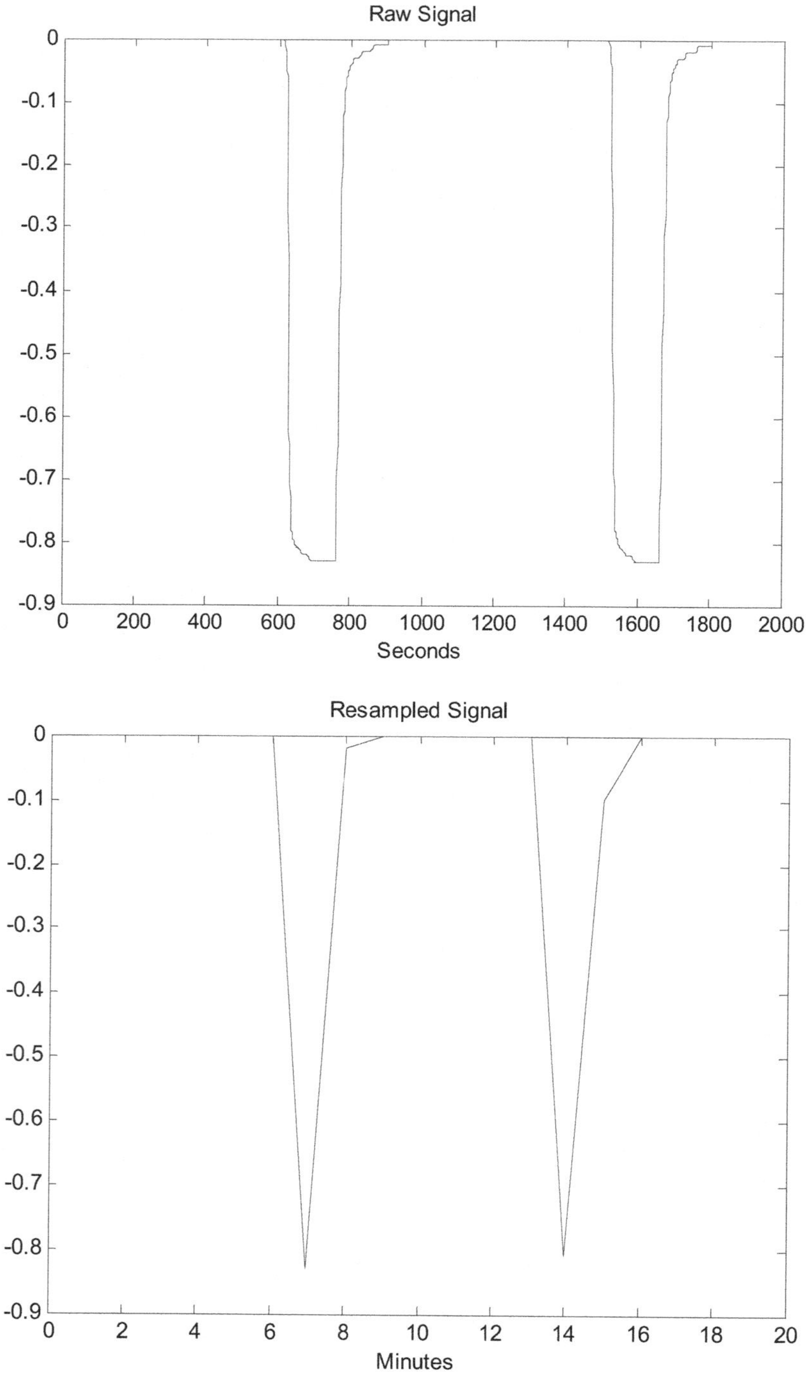

Figure 2.2 Output from the sensor simulator showing how the sensor's sampling interval can change the sensor output data. The top figure shows the continuous input signal and the bottom figure illustrates the signal changes that occur when a sensor has a discrete sampling interval of 1 sample per 2 minute interval.

Sensor Largest Measurement Value

The highest value a sensor can quantify or record can be specified as well. The value is given in parts per million (ppm) or milligrams per liter (mg/L). The dynamic range of each sensor ranges from zero to the largest measurement value. This parameter is used for expected noise and drift calculations as well as output data clipping when input values exceed the upper threshold.

Sensor Noise, Accuracy and Precision

At each sample point, a Gaussian distribution (e.g., a bell-shaped curve) is generated by the simulator that is centered on the input signal value. The published sensor accuracy and/or precision value influences the width of that distribution. The sensor noise terms reported in the vendor specification sheets are collected and grouped together to yield a total deviation or percent deviation for a particular measurement. The sensor reading that is passed on through the simulator at any particular point in time is randomly selected from this distribution. Selection of the output value from this distribution serves to introduce noise into the reading just as an operating sensor would. An illustration of how a noise-free input signal may be perturbed by a sensor with a high noise parameter is shown in Figure 2.3.

Cross Contaminants

A sensor can also respond to contaminants or interferents, other than the ones of interest, which may also be present in the water system. Response curves for these cross contaminants must be incorporated into the simulation so the sensor response can be taken into account. This feature is currently not implemented in the simulator since commercial sensor specifications do not usually include cross contaminant response information. Specialized testing is required to generate cross contamination or interference data and is beyond the scope of this study.

Sensor Drift

Several algorithm choices have also been implemented that simulate sensor baseline drift. Linear drift is a constant increase or decrease applied to the data over time. The degree of drift applied is taken directly from the sensor specification sheets. Sinusoidal drift requires an amplitude and period to be specified by the user in order to derive a varying baseline, which is added as an offset to the sensor data over time. Piecewise-linear drift consists of linear curves of varying slope and intercepts that are concatenated together over time. The incorporation of drift into the simulator is important for simple detection schemes that may involve a voltage threshold level. Sensor electronics drift may cause the sensor output to exceed a preset alarm threshold even when no contaminant is present in the sample. An example from the sensor simulator showing how sensor drift can alter an input signal is included in Figure 2.4.

Sensor Resolution

The resolution of the sensor identifies the smallest discernable change in the measured quantity. A sensor with high resolution will track changes in water quality with more accuracy and smoothness than one with low resolution. How the sensor resolution affects overall performance is also very much dependant on the sensitivity of the change detection algorithm that is in place to set an alarm condition. The resolution is specified in terms of significant digits to the right of the decimal place for reported units of residual chlorine or total organic carbon in milligrams per liter of water.

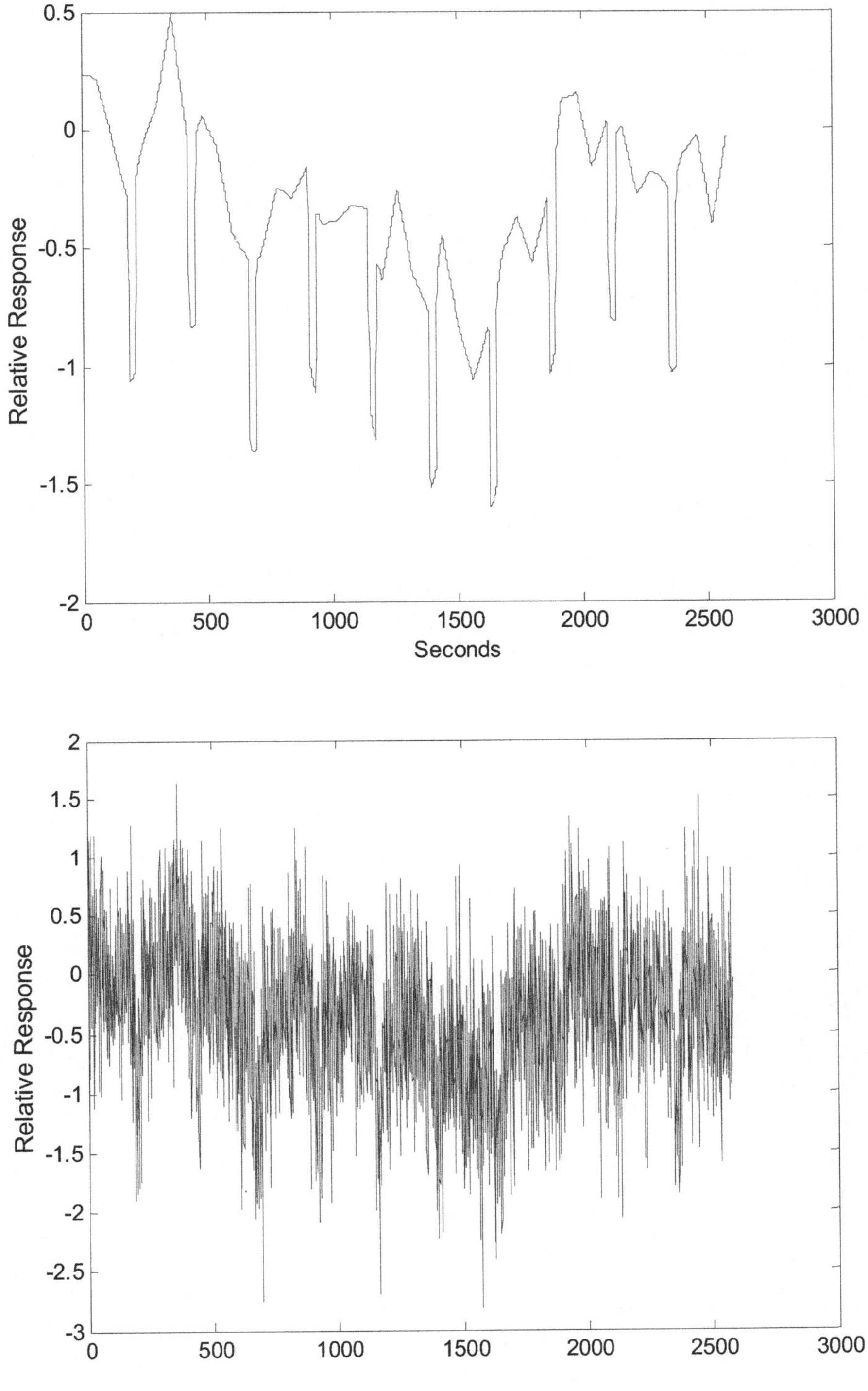

Figure 2.3 Output from the sensor simulator showing how sensor noise characteristics change the sensor output. The top figure shows the continuous noise-free input signal and the bottom figure illustrates the changes that occur in the sensor output for a relatively noisy sensor.

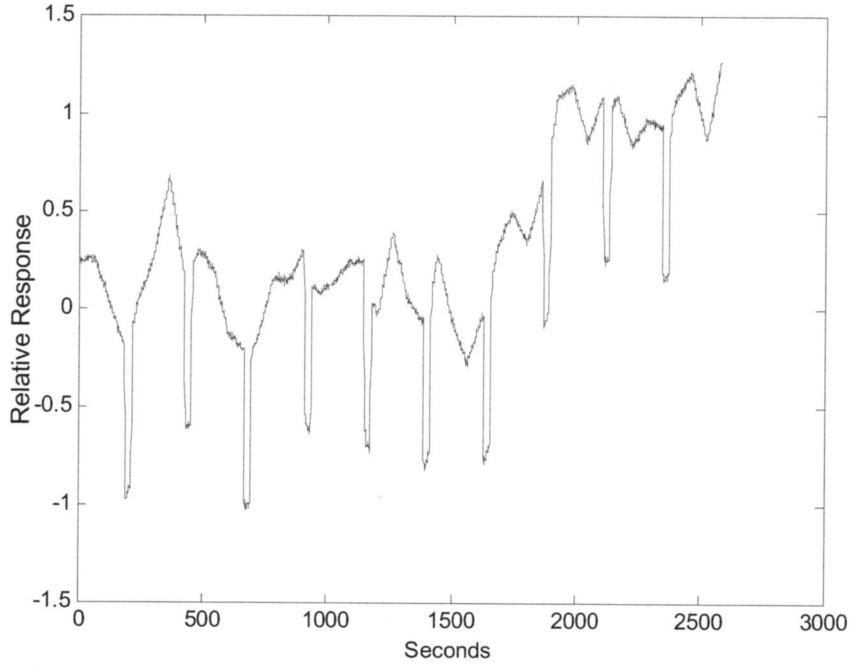

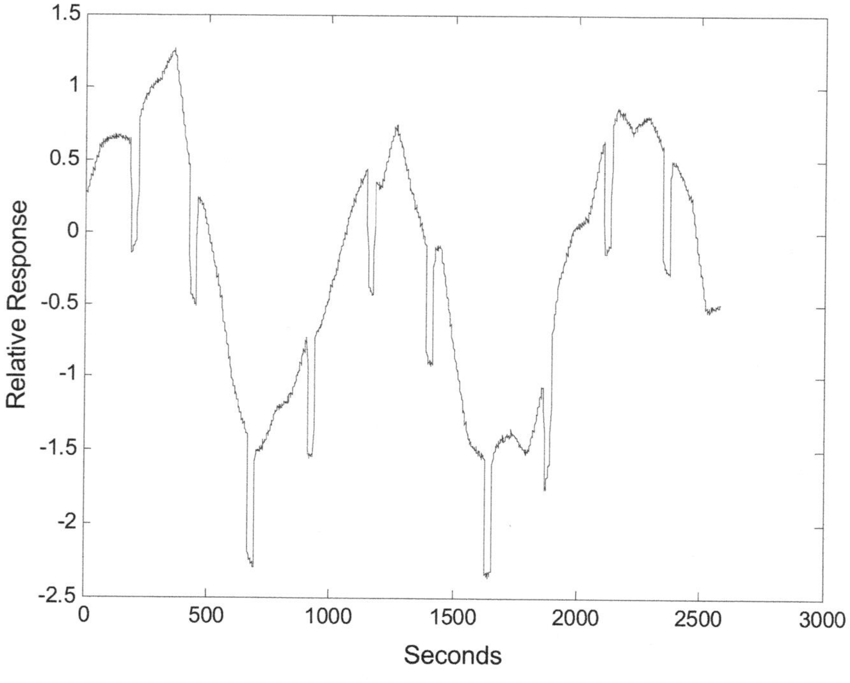

Figure 2.4 Output from the sensor simulator showing how sensor drift characteristics change the sensor output data. The top figure shows the continuous drift-free input signal and the bottom figure illustrates the changes that occur in the sensor output with a sinusoidal drift rate of 0.5 response units per 1000 seconds.

Contaminant Pulse Shape and Length

The simulator includes the ability to incorporate various contaminant pulse shapes for sensor evaluation. Contaminant pulse shapes are determined by the manner in which a particular contaminant is injected into a water system. A fast injection produces a plug of high-concentration contaminant moving through the system; whereas, a lower injection rate over an extended period of time will produce downstream pulse shapes that change more slowly and are less concentrated. In most simulation trials, the expectation is that contaminant pulse shapes will be determined through a separate modeling test using a hydraulic flow model such as EPANET that enables the pulse shape to be estimated at downstream locations following a specified injection scenario in a particular distribution layout. Additional information on how pulse shapes and pulse lengths were determined for this evaluation is given in following sections of this report.

Contaminant Pulse Spacing

The amount of time between contaminant pulses used in the simulation can be specified in seconds, minutes or hours. This feature within the simulator enables repeated testing of the same contaminant pulse form in the simulation run. The repeated pulses are used to develop true positive, false positive, true negative and false negative detection statistics for the designated sensor/contaminant configuration.

Water Quality Baselines

A number of baseline options to simulate typical fluctuations in measured water quantities within a water system are included in the simulator. The simulator includes a checkbox for inclusion of a user-supplied water quality baseline. If the baseline checkbox is set, a water quality baseline is added to the contaminant pulse prior to processing through the sensor simulator as illustrated previously in Figure 2.1. A synthetic baseline can also be used that is generated using a correlated time series as described in a recent paper on testing water quality change detection algorithms (McKenna et al. 2006). In this evaluation, only real baseline data from water utilities was used. Multiple weeks of continuous baseline data for total organic carbon were obtained from a large Eastern U.S. water utility (unnamed by request of the utility). Measurements were made at 1-min intervals using a UV probe calibrated for TOC response. Similar baseline data from residual chlorine sensors were obtained from a different intermediate-sized water utility in the Eastern U.S. Residual chlorine measurements were continuously made at 3-minute intervals. These data sources provided site-specific water quality baseline information that could then be further combined with contaminant peak information for further sensor/change detection system processing.

Sensor Simulator User Interface

A screenshot of the sensor simulation program's graphical user interface is shown in Figure 2.5. The sensor type and performance attributes, baseline choices, contaminant type and frequency characteristics are all input by the user on the left panel. The input signal is processed using the selected sensor performance parameters by pressing the execute button. The resulting

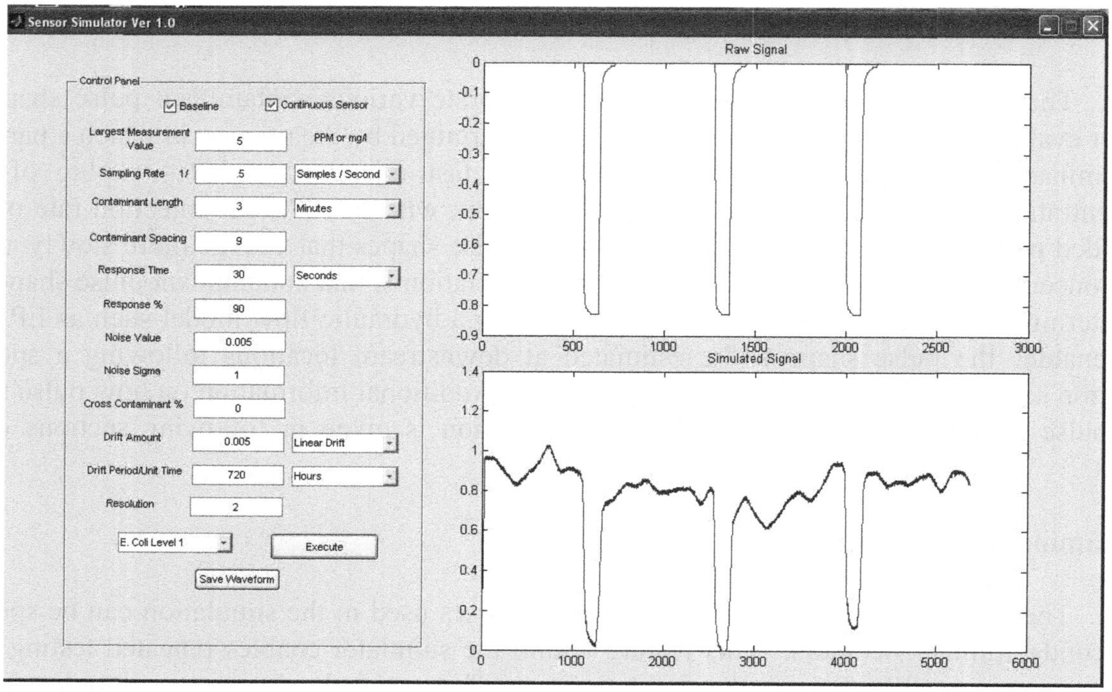

Figure 2.5 A screenshot of the graphical user interface for the sensor simulation software. The top trace is the raw input signal and the bottom trace is the sensor output with sensor noise and drift parameters applied.

signal degradation caused by the specified sensor performance attributes is shown on the right panel. The top plot shows the contaminant signal (sensor output with time) as it would appear in a perfectly noiseless system and the bottom plot is the simulated output response as processed through the sensor, where sensor noise, sampling rate, electronic drift and water quality baseline changes are all taken into account. In the example shown, the basic shape of the contaminant peak is preserved; however, the noise characteristics imposed by the imperfect sensor cause some baseline perturbation.

DESCRIPTION OF THE SIMULATION PROCESS

A schematic diagram of the entire simulation process is given in Figure 2.6 and is also summarized in the following list. Additional details concerning specific aspects of the simulation process are included in subsequent paragraphs.

- Conduct EPANET runs for short- and long-pulse contamination events using the pipeline layouts from a real water distribution system.
- Obtain contaminant pulse shape at selected locations downstream from the injection point using EPANET data outputs.
- Linearly scale the downstream pulse shape to a specific no-health-effect dose level for the target contaminants.
- Re-scale the observed contaminant concentration levels to both TOC and residual chlorine equivalents using published sensor response functions.

- Superimpose a series of re-scaled contaminant pulses onto the baseline data trace. The pulse data and baseline data are added together to produce a composite signal.
- Process this combined signal through the sensor simulator with sensor performance characteristics set for the sensor of interest (e.g., sampling rate, noise level, drift etc.).
- Iteratively process the output signal from the sensor simulator through the event change detection module. Vary the threshold value (i.e. the event change software "sensitivity setting") with each iteration and compile the detection statistics (e.g., true positive, false positive, true negative, false negative) for each threshold setting.
- Compile the data from the iterative runs into a format suitable for analysis using a Receiver Operating Characteristics (ROC) approach. Change detection results are tabulated for a given threshold setting in the form of true positive, false positives, false negative and true negative counts. Further analysis of ROC curve data serves as an indicator of overall sensor/algorithm performance.
- Repeat the simulations for other concentration levels, sensor types, contaminant pulse lengths, and baseline conditions.

Contamination Agents and Scenarios

Two chemical contaminants, aldicarb—a carbamate pesticide (2-methyl-2-(methylthio) propionaldehyde O-methylcarbamoyloxime) and sodium fluoroacetate—a mammalian pesticide, were chosen for study in this investigation. Rationale for selection of these two particular chemicals is based on the fact that these contaminants have been used in earlier EPA water contamination studies at the Cincinnati test loop facility and that residual chlorine and TOC sensor response data for these two substances have been published (Hall et al. 2007). Both of these chemical contaminants can be classed as moderately potent toxins with sodium fluoroacetate being the more potent of the two. Our research efforts were limited to the investigation of chemical contamination events only since evaluation of biological contamination by the surrogate sensors being considered here is dependant on the presence of bacterial growth medium. Response factors of the candidate residual chlorine and total organic carbon sensors are at present only qualitatively known for biological growth medium.

The contamination scenarios chosen can be described as follows. Chemical contaminant is injected directly into the distribution system in either a short (15-minute) or long (30-minute) interval. The amount of contaminant injected is scaled to yield selected dose levels at some arbitrary distance downstream from the injection point. (See following discussion on contaminant transport modeling and contaminant pulse height scaling carried out between the injection and sampling points.) The target dose level at the sampling point for the majority of simulation runs was at the No Observable Adverse Effect Level (NOAEL), which is the dose level that is expected to yield no observable adverse effects following chronic exposure to the toxin. The exposure is assumed to be consumption of a 250-ml volume (one glassful) of water. The NOAEL dose for a given chemical is based on the results of toxicological investigations in various animal models and/or epidemiological studies along with expert judgment to extrapolate these data to human dose-response effects (ATSDR 2007). Use of the NOAEL as the target dose is conservative in the sense that this dose level is indicative of no potential adverse health effects following chronic daily exposure. Thus, a single exposure at the NOAEL dose level is even more unlikely to result in observable adverse health effects when compared to a chronic exposure scenario. The NOAEL data for aldicarb and sodium fluoroacetate are given in

Table 2.4 (IRIS 2007a, IRIS 2007b). For comparison, the lowest observable adverse effect level (LOAEL), a higher dose level associated with observed toxic effects and the NOAEL doses are shown in the table along with equivalent in-pipe concentration for the assumed 250-mL water consumption.

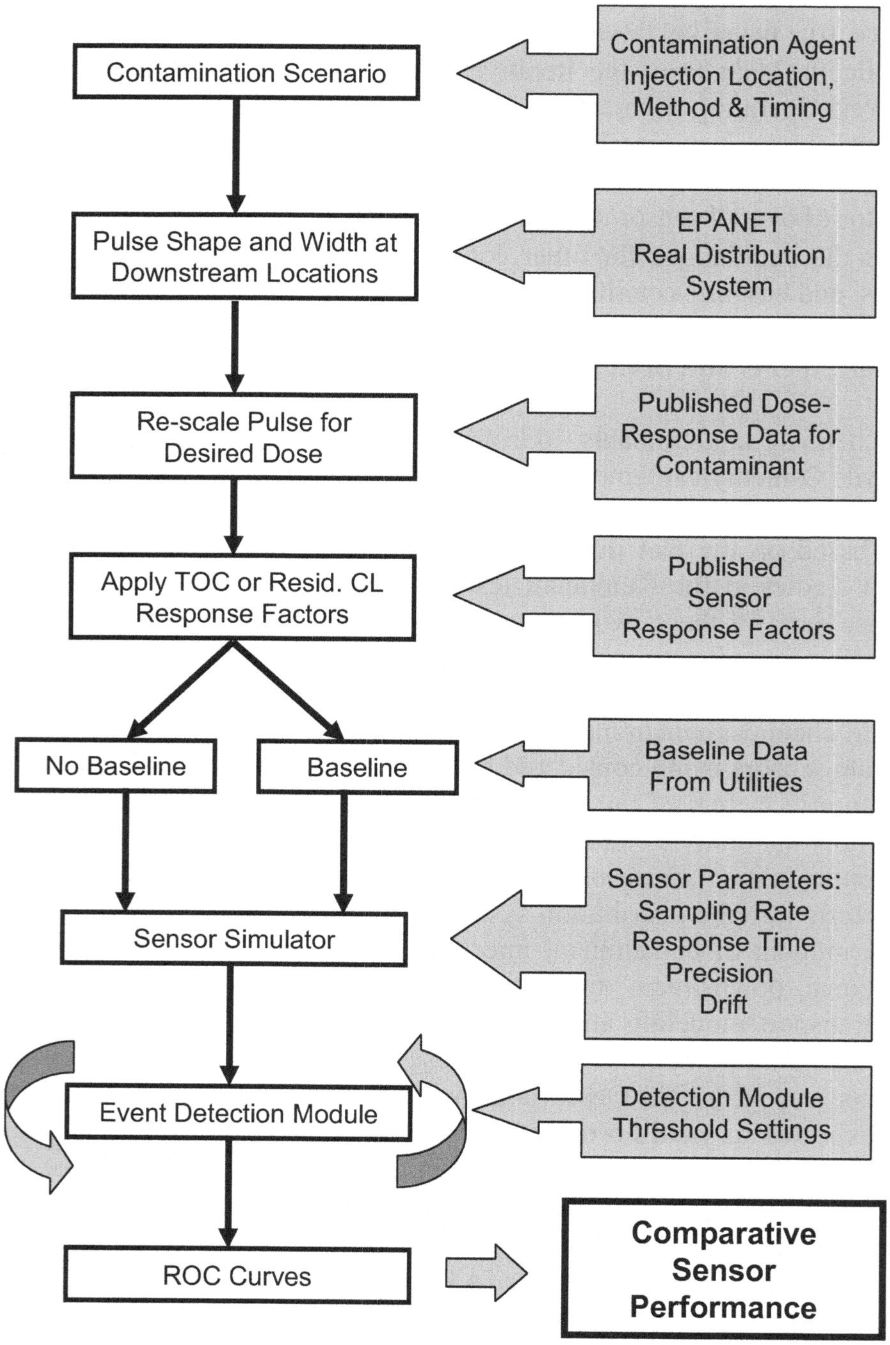

Figure 2.6 A flow diagram illustrating the sensor simulation process for a contamination event.

Table 2.4
Target Contaminant Dose Information and Equivalent In-pipe Concentrations

Contaminant	LOAEL mg/kg/day	Equiv In-pipe Conc. at LOAEL Dose (mg/L)	NOAEL mg/kg/day	Equiv In-pipe Conc. at NOAEL Dose (mg/L)
Aldicarb	0.58	162	0.11	31
Sodium Fluoroacetate	0.20	56	0.05	14

Distribution System and Hydraulic Flow Characteristics

We used the pipe network design from a relatively small water system (~3 million gallons per day) for the development of the contamination scenarios using EPANET. Developed by EPA's Water Supply and Water Resources Division, EPANET is software tool that models water flow through distribution piping systems. EPANET is a Windows-based program that performs extended-period simulation of the hydraulic and water quality behavior within pressurized pipe networks. Pipe networks consist of pipes, nodes (pipe junctions), pumps, valves, and storage tanks or reservoirs. EPANET tracks the flow of water in each pipe, the pressure at each node, the height of the water in each storage tank, and the concentration of a chemical species throughout the network during a simulation period (Rossman 2000, EPA 2007a). Key attributes of the distribution system are given in Table 2.5 and a diagram of the pipe network is shown in Figure 2.7.

Contaminant Pulse Characteristics

The contaminant pulse associated with a short (15-min) and long (30-minute) injection period was simulated using the previously described distribution system using the EPANET model. Typical examples of the modeled contamination events are given in Figures 2.8 and 2.9. The transport and dilution of a unit measure of injected contaminant was modeled through the system and the shape of the contaminant pulse was determined at several downstream nodes or locations. The distance from injection point to measurement point for this simulation was about 900 feet and was arbitrarily chosen. The original square-wave contaminant pulse shape of the contaminant is moderated following transit through the pipe network from the injection point to the downstream node selected for study. The transport of the contaminant from the injection point to the sensor location is modeled as a non-reactive tracer within EPANET.

Table 2.5
Summary Description of Distribution System Used in Simulations

Parameter	Value
Distribution System	534 Pipes, 394 Junctions, 2 Storage Tanks, 4 Valves, Gravity Feed Only
Contaminant injection rate	18.9 L/min (5 gallons/minute)
Injection concentration	500 mg/L
Injection duration – Long	30 minutes
Injection duration – Short	15 minutes
Distance from injection point to downstream node used as the sampling point	902 ft

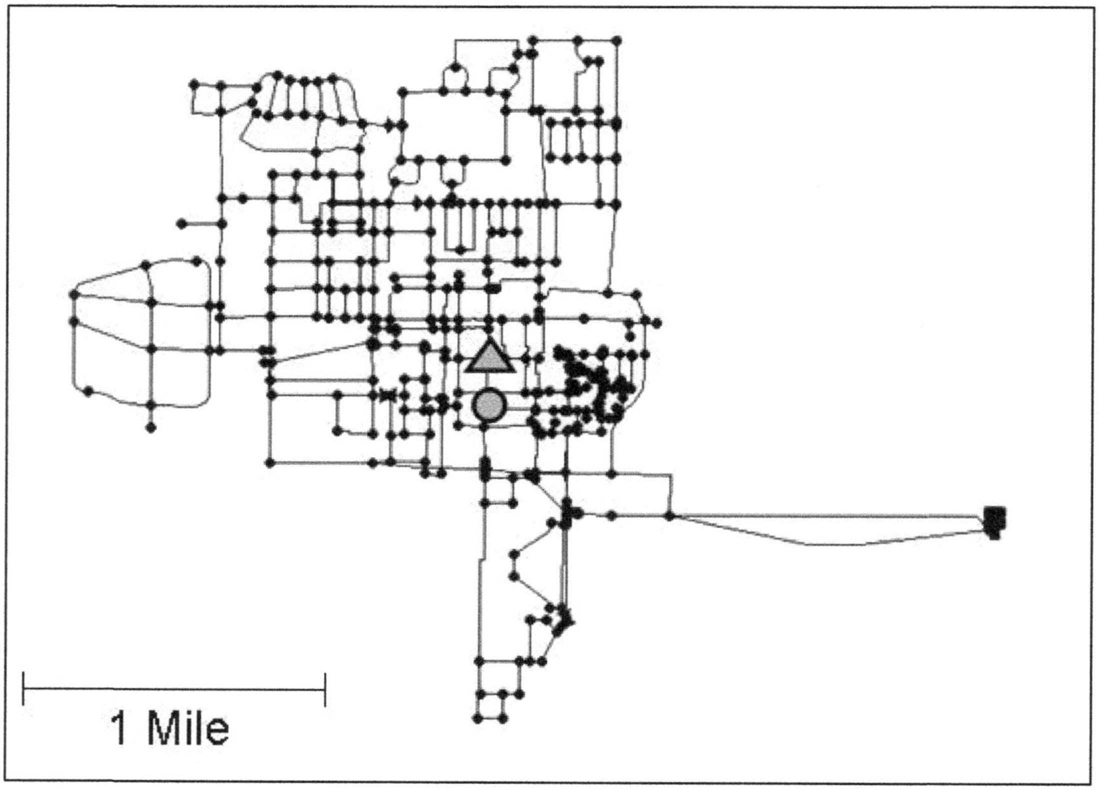

Figure 2.7 Schematic diagram of the water distribution system used in the simulations. The injection point is near the center of the figure, denoted by the grey circle. The downstream pipe junction where the measurements were simulated was 902 feet distant from the injection point as indicated by the triangle.

Converting the Contamination Pulse to Equivalent Sensor Response

We used sensor response data published by Byer and Carlson (2005) to estimate the increase in total organic carbon and the reduction in residual chlorine signal that would be observed following exposure to a NOAEL dose concentration level in the pipe for each of the two target contaminants. Expected TOC and residual sensor response to varying concentrations of aldicarb and sodium fluoroacetate are shown in Figures 2.10-2.13. The plots show a linear response and the slope of the line, calculated by linear regression, is the specific response for each sensor type (e.g., change in TOC or residual chlorine reading per unit change of contaminant concentration). The sensor response factors for the two contaminants are summarized in Table 2.6. For this calculation we assume a similar response of all sensor types to the contaminants of interest.

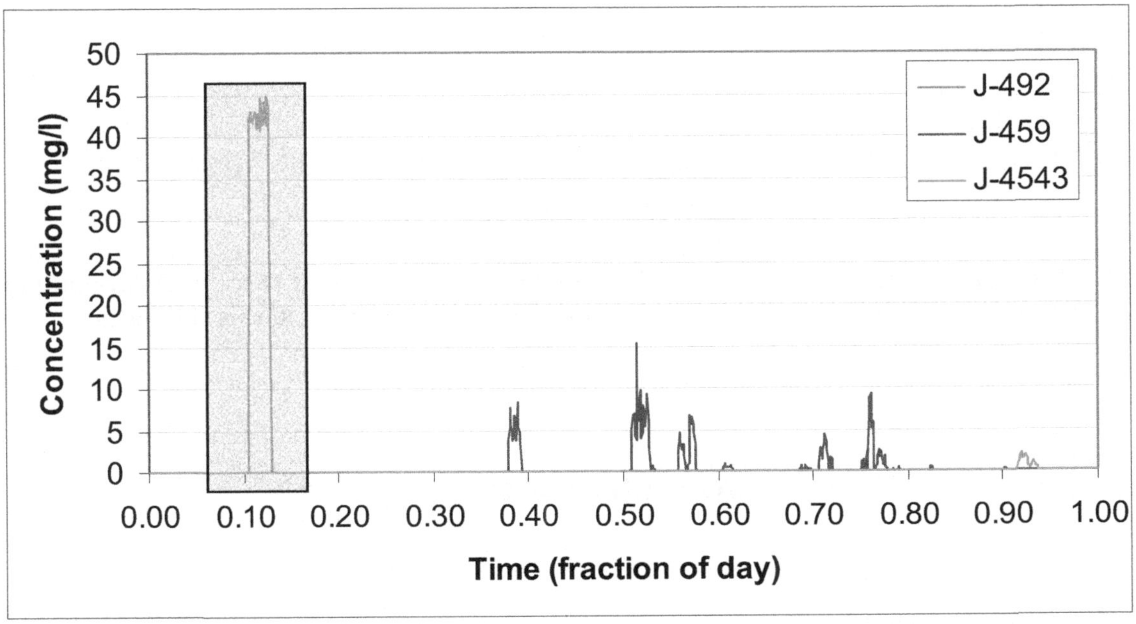

Figure 2.8 EPANET results for a long (30-min) contaminant injection at three downstream pipe junctions. The highlighted box denotes the pipe junction selected for study in this simulation.

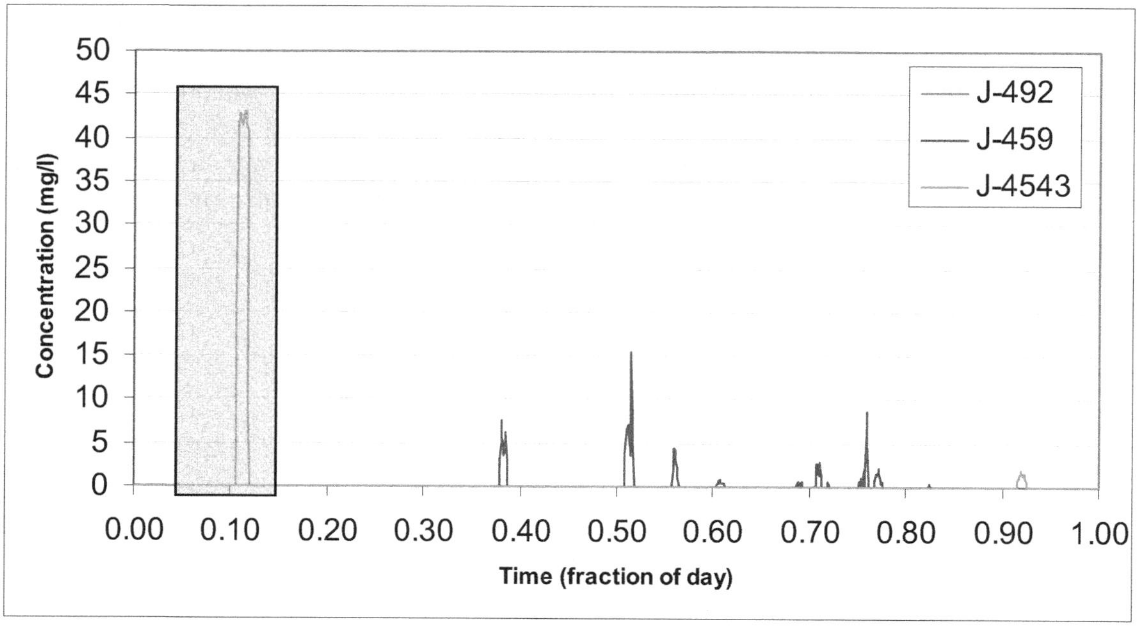

Figure 2.9 EPANET results for a short (15-min) contaminant injection at three downstream pipe junctions. The highlighted box denotes the pipe junction selected for study in this simulation.

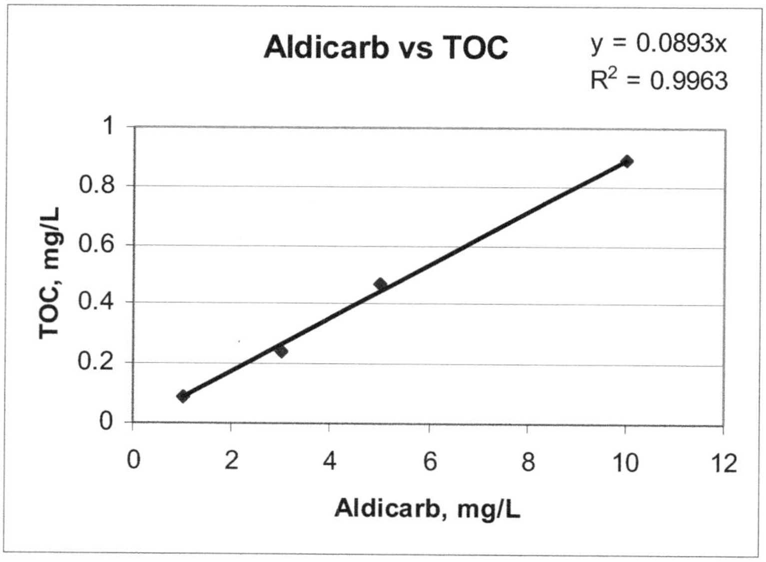

Figure 2.10 Expected total organic carbon sensor response to aldicarb

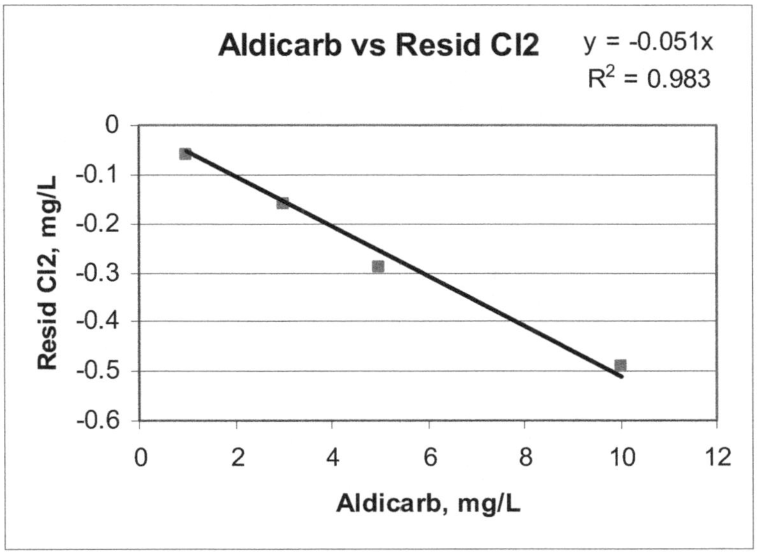

Figure 2.11 Expected residual chlorine sensor response to aldicarb. Note that a decrease in background residual chlorine levels are expected during a contamination event.

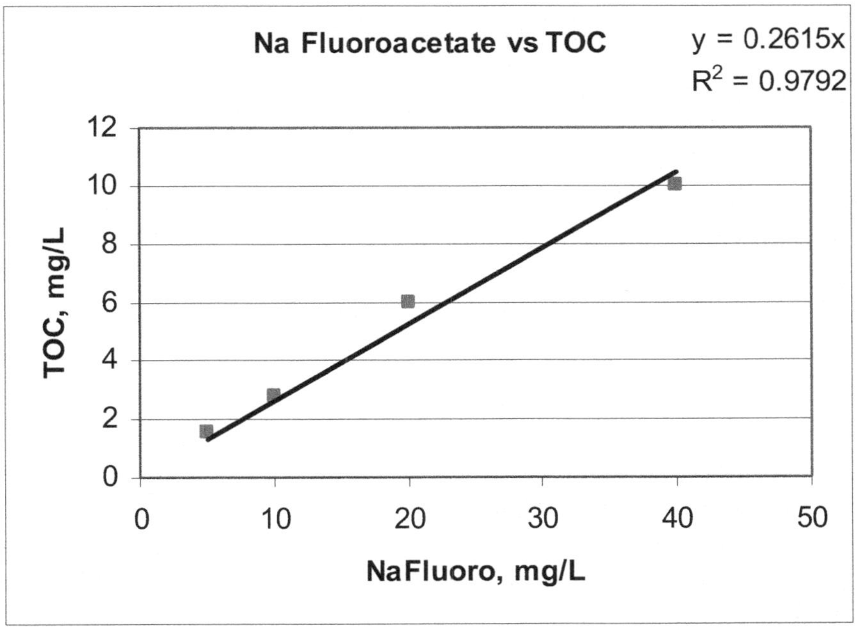

Figure 2.12 Expected total organic carbon sensor response to sodium fluoroacetate.

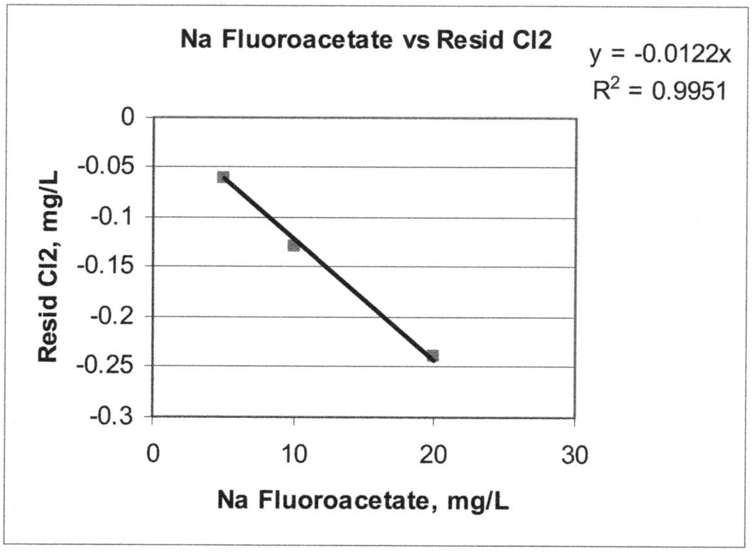

Figure 2.13 Expected residual chlorine sensor response to sodium fluoroacetate.

Table 2.6
Response Factors for Total Organic Carbon and Residual Chlorine Sensors

Contaminant	Total Organic Carbon	Residual Chlorine
Aldicarb	0.089	-0.051
Sodium Fluoroacetate	0.262	-0.012

For determining the expected TOC sensor reading at the aldicarb or sodium fluoroacetate NOAEL dose the following expression was used:

$$E = \frac{D \bullet W}{V} \bullet S \qquad (1)$$

E = expected TOC sensor reading at NOAEL dose level (mg/L)
D= NOAEL (mg/kg/day)
W= 70 (kg) avg male weight
V= 0.25 (L) water consumed
S = sensor specific response factor (mg/L TOC per mg/L aldicarb or sodium fluoroacetate)

The same expression was used to calculate the expected residual chlorine sensor response, with the understanding that the residual chlorine response moves in a negative direction.

Scaling the Contamination Pulse to a Dose Level

The concentration pulses shown in Figures 2.8 and 2.9 were scaled to the NOAEL dose levels for aldicarb and sodium fluoroacetate, as determined by the previous equation. The following scaling equation was used:

$$C_{Scaled} = C \cdot (E/[C_{mean}]) \qquad (2)$$

where C is the concentration level in the original concentration pulse, E is the expected signal corresponding to the NOAEL concentration level (derived in the previous equation) and C_{mean} is the average of the original concentration pulse (from the point of baseline departure of the signal to the return to baseline). The scaled NOAEL, 15-minute TOC pulse for aldicarb is shown in Figure 2.14 and the scaled NOAEL 15-minute TOC pulse for sodium fluoroacetate is shown in Figure 2.15. Similar scaled plots for the residual chlorine signal are shown in Figures 2.16 and 2.17. As previously noted, in the case of residual chlorine, the baseline undergoes a negative deflection during contaminant exposure since the presence of the chemical represents a chlorine demand which consumes some portion of the residual chlorine present in the treated water.

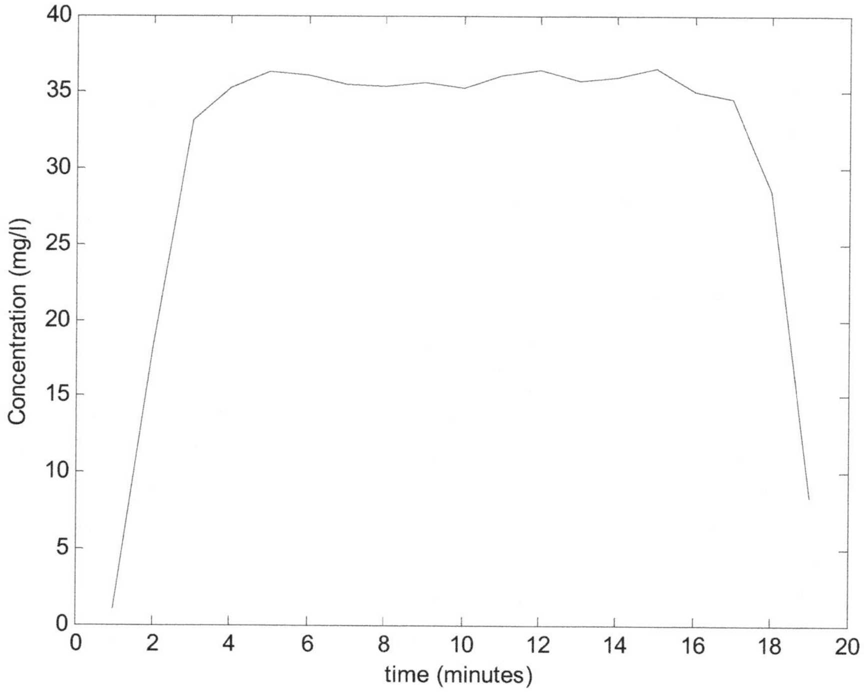

Figure 2.14 The downstream 15-minute TOC signal scaled to the aldicarb NOAEL dose.

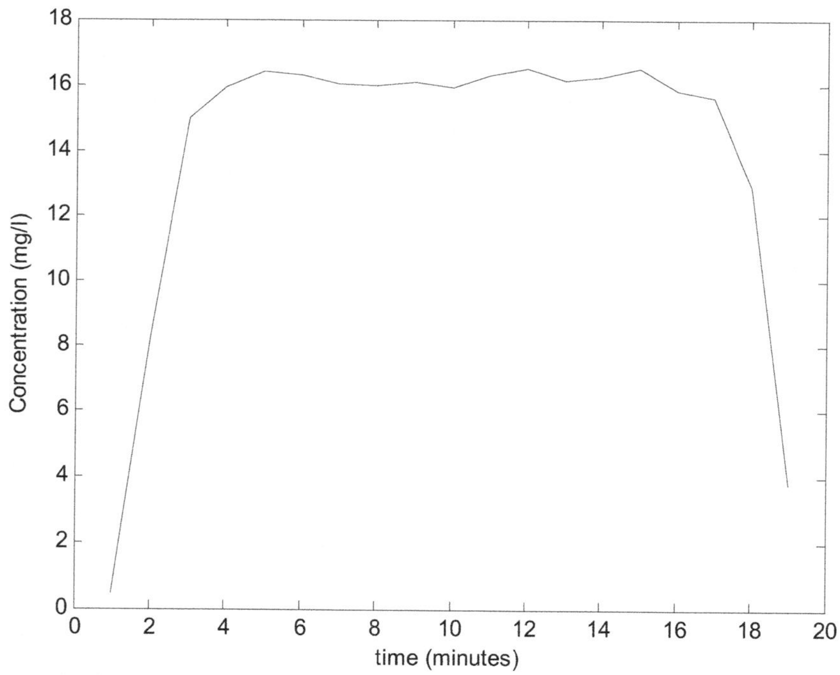

Figure 2.15 The downstream TOC signal scaled to the 15-minute, sodium fluoroacetate NOAEL dose.

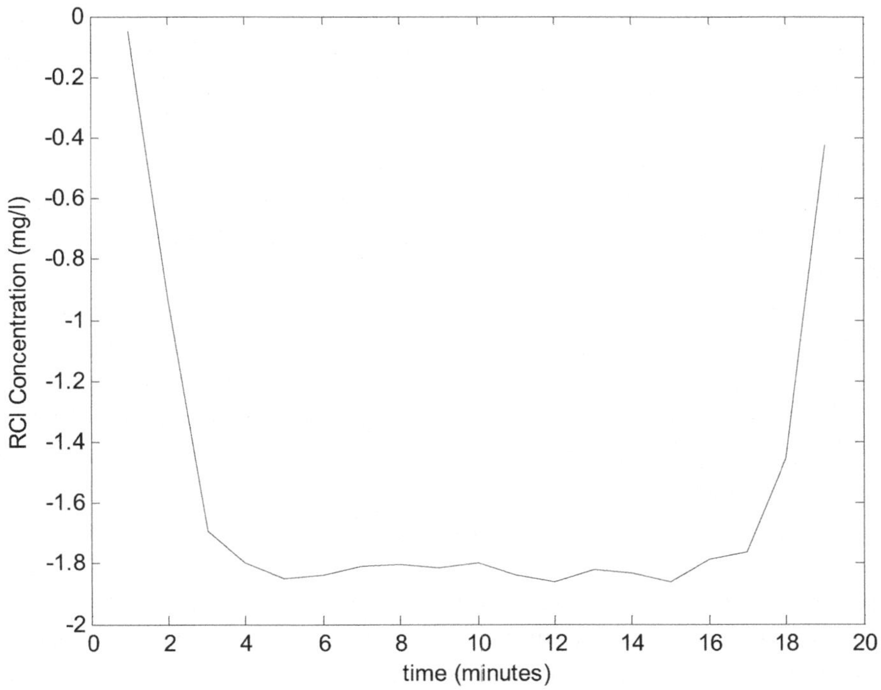

Figure 2.16 The downstream chlorine signal scaled to the 15-minute, sodium aldicarb NOAEL dose.

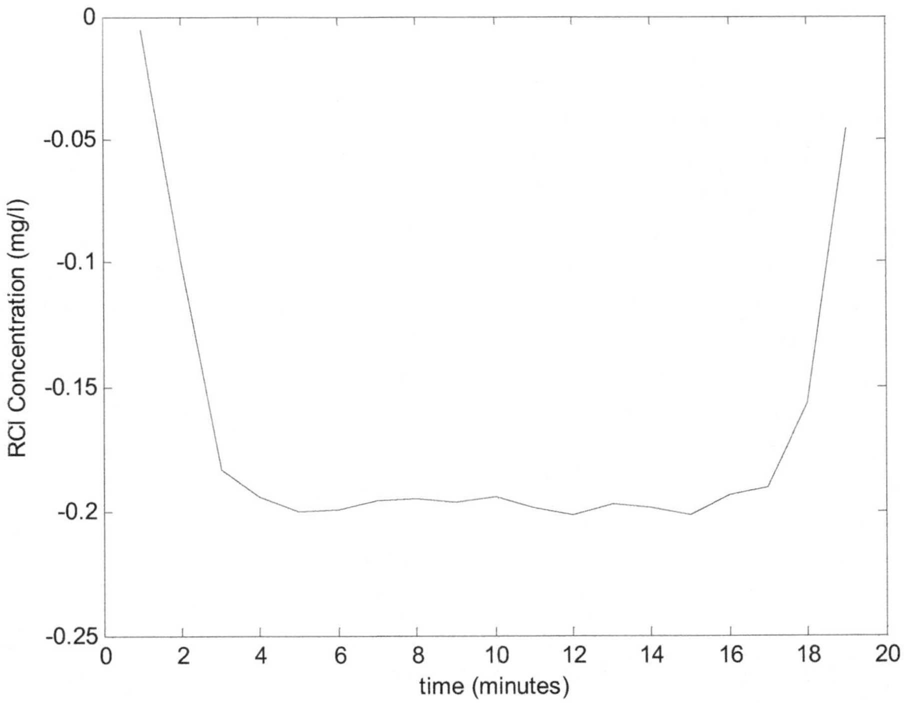

Figure 2.17 The downstream residual chlorine signal scaled to the 15-minute, sodium fluoroacetate NOAEL dose.

Inclusion of Water Quality Baselines

As a part of this research, several actual water quality baselines for both residual chlorine and total organic carbon were obtained from selected US water utilities and were incorporated into the simulation. A residual chlorine baseline, derived from on-line monitoring at a large metropolitan utility is used for testing residual chlorine sensors. Similarly, a total organic carbon baseline is included for testing TOC monitors. In the interest of security, no further attribution is given for these data sources. In each case a baseline data set that spans multiple days is included within the simulator.

An example of TOC-equivalent baseline measured with a UV probe in a utility's distribution water system is shown in Figure 2.18. A similar example of a residual chlorine baseline is given in Figure 2.19. All baseline fluctuations are deemed "normal" and are representative of the types of variability encountered in most utility measurement data. Water quality baseline changes over time may be attributable to any number of factors such temperature or water flow rate changes, local pipe maintenance activity, or water composition caused by changes in the mixing ratio of various source water supplies (e.g. surface and ground water) that are feeding the pipe being monitored.

It is important to note that the baselines shown do not strictly represent a true in-pipe signal in that they are too perturbed by the sensor system used to generate them. The data have been smoothed to reduce sensor-induced noise and the assumption is made that variations noted are attributable to changes in water quality and not induced by the sensor itself. This may not be true for drift; however, the time interval of the baseline used in the actual simulation run is very short and essentially free from the influence of baseline drift.

Event Detection Algorithms

The altered signal is output from the sensor simulator and further processed through one of three event detection algorithms. The first and simplest method is a threshold setting that corresponds to a comparison of the signal at any particular point in time with a preset threshold level. Excursion of the contaminant signal above this threshold, in the case of TOC, and below this threshold, in the case of residual chlorine, results in the detection of a true positive "event." Conversely, failure of a contaminant signal to exceed the threshold results in a non-event (false negative). The contamination pulses with and without added water quality baselines are repeatedly cycled through the simulator as threshold settings are varied in order to count the number of true positives and false negatives for each threshold settting. The threshold settings are varied through the iterations as follows: The entire baseline signal (without pulses added) is processed to determine the minimum (S_{min}) and maximum (S_{max}) values. The threshold settings are then varied in increments of 0.1 mg/L for both TOC and residual chlorine from a starting value $S_{min} - 0.1$ to $S_{max} + 0.1$ mg/L.

The other two event detection algorithms have been developed within the EPA-sponsored Threat Ensemble Vulnerability Analysis (TEVA) project and employ additional statistical analysis of the sensor data stream. These routines are formally implemented in the CANARY software package (Hart et al. 2007). The two methods are termed forward linear predictor (FLP) and multivariate distance (MV).

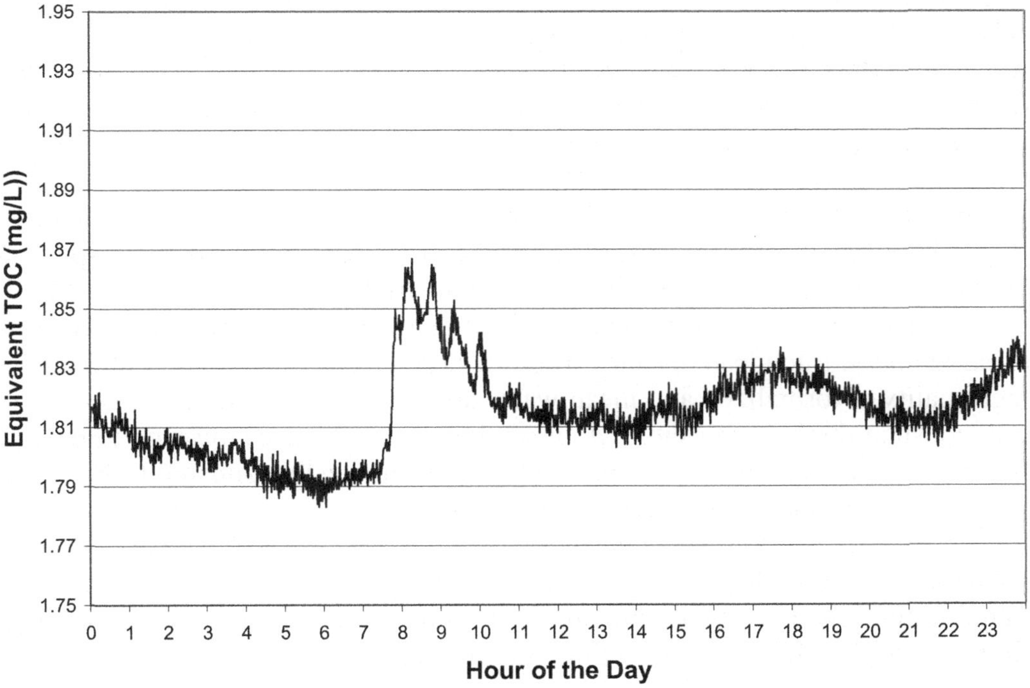

Figure 2.18 TOC-equivalent baseline from a metropolitan utility distribution line as measured by a UV-visible probe. Measurements are made in one-minute intervals. This is an example of baseline data that are merged with contaminant peak data for the sensor performance simulation.

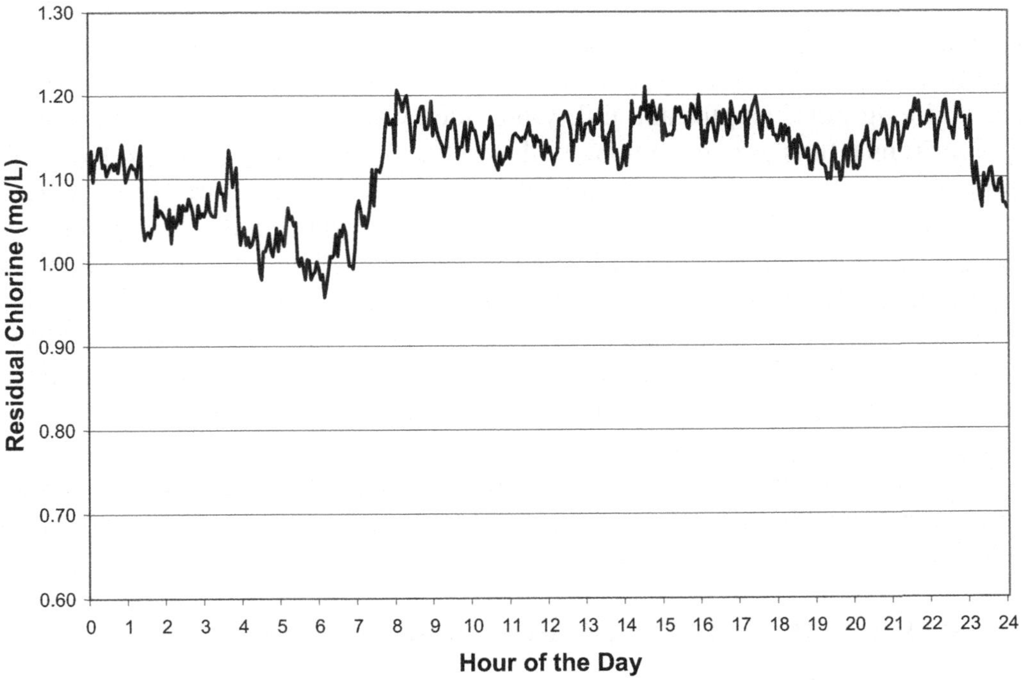

Figure 2.19 Residual chlorine baseline from a metropolitan utility distribution line as measured by a chlorine sensor. Measurements are made at three-minute intervals. This is an example of baseline data that is merged with contaminant peak data for the sensor performance simulation.

The linear predictor (LP) algorithm estimates the current value of a time series based on past samples. In its most general form, this approach is also known as the autoregressive moving average predictor (Bras and Rodriguez-Iturbe 1993). Statistical means are used to compare the observed value with the estimated value and decide whether the discrepancy represents an expected value (no event) or an anomaly (event). Note here that each individual time step can be considered a potential event.

The multivariate distance (MV) algorithm considers changes in water quality based on the difference between successive water quality measurements in multivariate space (Klise and McKenna 2006). In these simulations the data are univariate so much of the power of the routine is not utilized. The distance measure used to distinguish between normal and anomalous water quality is based on the Euclidean distance from the normalized water quality measurement to the mean water quality. Normalized and mean water quality is a function of the size of the moving window. A constant distance threshold, d_c, is applied to the distance measure to determine if the water quality measurement is anomalous. Distances above the threshold are categorized as anomalous, while distances below the threshold are considered normal. More complete descriptions of the FLP and MV event detection methods are included in Appendix A.

As with the simple threshold method, the detection sensitivity is variable with the FLP and MV algorithms as well. The output signal from the sensor simulation module is iteratively processed through the threshold detection algorithm with an accompanying change in the detection sensitivity setting with each pass through the routine. Twenty five iterations are made through the routine and the resulting compilation of either detects (true positive) and non-detects (false positives) are compiled and used for additional analysis, as further described in Chapter 3.

SIMULATION MATRIX SUMMARY

A summary of all simulations completed in this study is given in Table 2.7. For each of four TOC sensor types, four runs (Runs 1-4) were done with aldicarb at the NOAEL dose level with short- and long-pulse lengths and utility-provided water quality baseline inclusion on or off. Similarly, four runs (Runs 5-8) were done with sodium fluoroacetate at the NOAEL. In addition four additional runs (Runs 9-12) were done at an aldicarb dose that was about 4 percent of the NOAEL dose level to simulate how well a CWS system would detect compounds with higher toxicity levels. Two more runs (Runs 14 and 15) were carried out at a dose level that was about 0.2% of the aldicarb NOAEL dose to simulate how well the sensor systems could detect highly toxic materials. Finally, a single run (Run 13) was done with no superimposed contaminant peaks in order to assess false positive rates that would likely be encountered at typical threshold settings for the three change detection algorithms. Additional description of Run 13 is given in the following section.

A total of 90 simulation runs were completed for the four TOC sensor types evaluated. Similar runs were carried out for two residual chlorine sensors for a total of 30 simulations for residual chlorine sensor performance evaluation.

FALSE POSITIVE SIMULATION METHOD DESCRIPTION

In order to investigate how the sensors and algorithm perform under baseline conditions with no added contaminant pulses, an additional simulation run (Run 13, as shown in Table 2.7) was executed to derive a measure of false positive rates for specific sensor types. The simulation was set up as follows: The optimal threshold setting—that setting providing the highest ratio of true positive events to false positive events—for each of the three change detection algorithms was chosen from Run 5 (high-dose sodium fluoroacetate, baseline included, 30-min pulse) and Run 9 (low-dose aldicarb, baseline included, 30 minute pulse) by choosing the optimum threshold setting for each of the event detection schemes from the data analysis (further described in Chapter 3). These threshold settings were then set within each event detection method and simulations were processed through the sensor and event detection modules. In this case, a baseline signal without contaminant pulses was processed through the algorithms and the number of false positive and true negative events was tabulated. The resulting false positive rate is the ratio of the false positive count to the sum of the false positive and true negative counts in each simulation. This statistic provides another measure of the false alarm rate that can be used in addition to other operating statistics such as receiver operating characteristic curves, further described in Chapter 3, to assess overall sensor performance.

Table 2.7
Simulation Run Descriptions

Run No.	Contaminant	In-Pipe Conc. (mg/L)	Pulse Length	Baseline
1	Aldicarb	31	30-minute	On
2	Aldicarb	31	15-minute	On
3	Aldicarb	31	30-minute	Off
4	Aldicarb	31	15-minute	Off
5	Sodium Fluoroacetate	14	30-minute	On
6	Sodium Fluoroacetate	14	15-minute	On
7	Sodium Fluoroacetate	14	30-minute	Off
8	Sodium Fluoroacetate	14	15-minute	Off
9	Aldicarb	1.2	30-minute	On
10	Aldicarb	1.2	15-minute	On
11	Aldicarb	1.2	30-minute	Off
12	Aldicarb	1.2	15-minute	Off
13#	None	N/A	N/A	On
14	Aldicarb	0.05	30-minute	On
15	Aldicarb	0.05	15-minute	On

Note: Run 13 was used to assess false alarm rates for baseline data only with no contaminant Injection. Detection threshold was fixed at an optimal level for each of the three change detection algorithms examined in this study.

CHAPTER 3
RESULTS AND DISCUSSION

RECEIVER OPERATING CHARACTERISTIC METHODOLOGY DESCRIPTION

The receiver operating characteristic (ROC) approach was used as an analysis tool to assess and compare overall sensor performance in our study. The ROC concept was developed during World War II to improve the performance of radar detection algorithms. Since that time the approach has been further developed and is used extensively in the medical diagnostics field to optimize the interpretation of diagnostic tests (Swets 1988, Spackman, 1989, Green 1996).

An example of ROC data in graphical format is shown in Figure 3.1. The so-called ROC curve is a graphical representation of the frequency of true detects versus the frequency of false positives over the variable threshold range of the particular sensor/event detection module being evaluated. For each simulation run and at a specific threshold setting, a contaminant peak is either detected (true positive) or missed (false negative). Additionally, the sensor/event detection module may detect a peak when in fact there is not one present (false positive) or correctly report a no-detect (true negative). At each event detection module threshold setting, the occurrence of any of these four test results are compiled using a sensor data stream that has contaminant peaks interspersed with sections of no peaks. Water quality baselines may be added to the sensor signal or not, depending on the desired characteristics of the simulation run (as noted previously in Table 2.7). To generate all the data needed for a ROC curve, the threshold setting of the event detection module is incremented and the run is repeated until detection statistics are compiled over the threshold range of the event detection algorithm. These data are then plotted in a ROC curve format where the x-axis is defined as the probability of a false positive (also labeled false positive rate or false alarm rate) and the y-axis is the probability of a true positive (also labeled true positive rate or hit rate). These two ROC curve parameters are further defined as follows:

$$\text{Probability of TP or Hit Rate or } TPR = \frac{TP}{TP + FN} \qquad (3)$$

$$\text{Probability of FP or False Alarm Rate or } FPR = \frac{FP}{FP + TN} \qquad (4)$$

where TPR is the true positive rate, TP is the true positive count and FN is the false negative count for a particular threshold setting. Similarly, FPR is the false positive rate, FP is the false positive count and TN is the true negative count.

The information contained in the ROC curves is particularly useful in the design and configuration of a continuous monitoring system. Operators can gain a comparative understanding of the overall sensor/event detection module options and select sensor systems and alarm sensitivity thresholds based on their tolerance levels for false negatives and false positives. For example, assume the ROC curve shown in Figure 3.1 applies to a particular sensor and change detection algorithm combination that is under consideration for installation at a utility. For the purposes of illustration, we further assume that utility operators decide they are willing to tolerate a false positive rate of 0.1 (e.g., 10% of all alarm events are false positives—in reality, an unrealistically high level). Moving vertically from the 0.1 point on the x-axis to its intersection with the ROC curve yields a probability of about 0.6 for true positives on the y-axis. Thus, for this particular sensor configuration, water quality baseline influences and change detection algorithm sensitivity setting, every alarm event carries with it a 10% chance that it is a false positive and a 60% chance it is a true positive. Inter-comparison of various sensors systems with differing performance characteristics can be made using ROC curve shapes. The shape of the ROC curve describes the overall system performance in measures that are particularly important to utility operators, namely, the utility tolerance level for false positive and false negative events. Ideally, the ROC curve should be a trace that moves vertically from the zero-level on the x-axis, making a sharp bend to the right near the top of the y-axis.

An additional statistic that can be derived from the ROC curve analysis is the area under the curve or AUC. This value can serve as an overall indicator of the ability of a particular sensor/event detection module to accurately detect contamination events and minimize false positive events. An AUC value of 1 signifies a perfect system that has a true positive detection rate of 1 and a false positive detection rate of zero. An AUC value of 0.5 corresponds to the diagonal line shown in Figure 3.1. This represents a useless sensor/event detection combination since at any threshold setting the true positive rate and false positive rate are the same. (A coin toss would yield equivalent results.) Values of the AUC between 0.5 and 1 give a measure of the overall effectiveness of a sensor/event detection combination. Values close to 0.5 indicate poorer performance whereas values close to 1 reveal better sensor/event detection module performance.

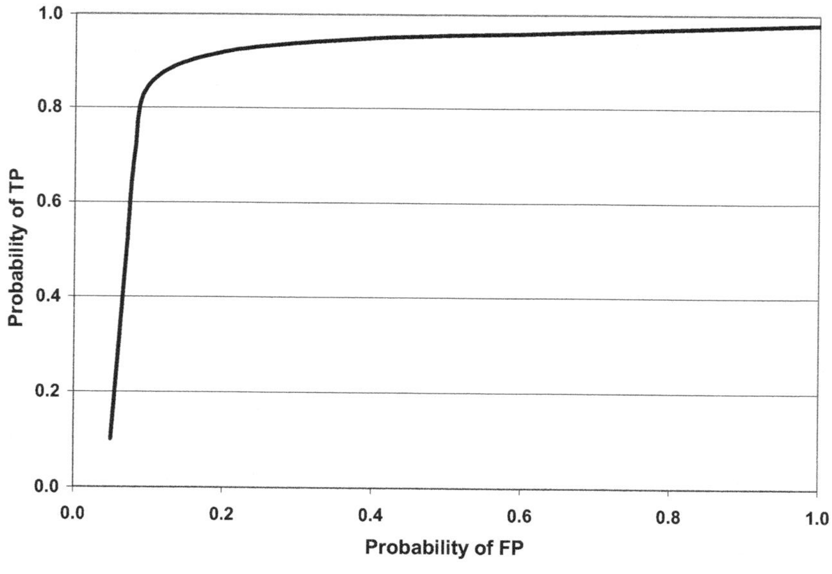

Figure 3.1 Example of a Receiver Operating Characteristic (ROC) curve.

SIMULATION RESULTS

Contaminant Peak and Baseline Data From Sensor Simulator

Several examples of the data streams from the sensor simulator to the change detection algorithms are shown in Figures 3.2 through 3.5. In the case of chlorine sensors the contaminant peak can be seen as a negative excursion below the added water quality baseline. In the first example shown for residual chlorine sensors (Figure 3.2), the contaminant peaks of a sodium fluoroacetate NOAEL dose level can be visually distinguished from the baseline data. The same is true for the second example (Figure 3.3) which is at 4 percent of a NOAEL dose level of aldicarb. The third example (Figure 3.4) is for a TOC sensor at an aldicarb NOAEL dose level and again, the contaminant peaks can be visually distinguished from the baseline. The final example (Figure 3.5) shows a sensor data trace at 4 percent of the aldicarb NOAEL level. In this case, the contaminant peaks cannot be visually distinguished from the baseline noise. A complete complilation of all sensor data output streams for the 15 simulation runs (as presented previously in Table 2.7) is included in Appendix B.

FLP Change Detection Algorithm Results

Computed results are categorized by algorithm type, sensor type and run number as previously described in Chapter 2. An example of a ROC curve for the two chlorine sensors evaluated in this study for Run 10 (Low aldicarb dose, baseline on, 15-min pulse) is shown in Figure 3.6. The Endress Hauser chlorine sensor shows better overall performance than the CLX sensor as evidenced by the closer proximity of the curve to the upper left hand corner of the plot. A typical ROC curve plot for the four TOC sensors is shown in Figure 3.7. The Endress Hauser and Spectro sensors show better performance with plots that more closely approach the upper left hand corner of the figure. The ROC curve areas for the forward linear predictor (FLP) algorithm are given in Table 3.1 for all 15 runs. The residual chlorine sensor results are un-shaded in the table and the total organic carbon sensors are shaded. The averages of all runs with and without baseline and with baseline only are also included in the table. For the residual chlorine sensors the EHCCS141 had the largest overall average with a ROC area of 0.88 and the largest baseline average with an AUC of 0.75. Bar graphs of ROC areas for chlorine and TOC sensors are given in Figures 3.8 and 3.9 respectively.

Overall, both residual chlorine sensors performed relatively well for the high (NOAEL) dose levels of aldicarb and sodium fluoroacetate with AUC values ranging from 0.73 to 0.97. The EHCCS141 performs considerably better then the CLX for the sodium fluoroacetate runs with baseline included. Both of the instruments show a decrease in detection performance for the low dose aldicarb runs with or without baseline. In this case AUC values range from 0.67 to 0.96 with the ESCCS141 showing better overall performance.

The TOC sensor data shows a high degree of comparability among the four sensors. The AUC values range from 0.73 to 0.98 for the high dose trials and, on average, the EHCCS70 performs best with an overall ROC average of 0.79 and baseline-only average of 0.80. A decrease in detection performance for all sensors is observed for the low-dose aldicarb runs with AUC values areas ranging from 0.53 to 0.83. Performance is similarly mediocre for the two very low dose trials with baseline included. The AUC values are near 0.5 for all sensors. Under the contamination scenario conditions chosen for this study, it appears that the combined sensor and algorithms are not useful for identifying these very low contamination levels.

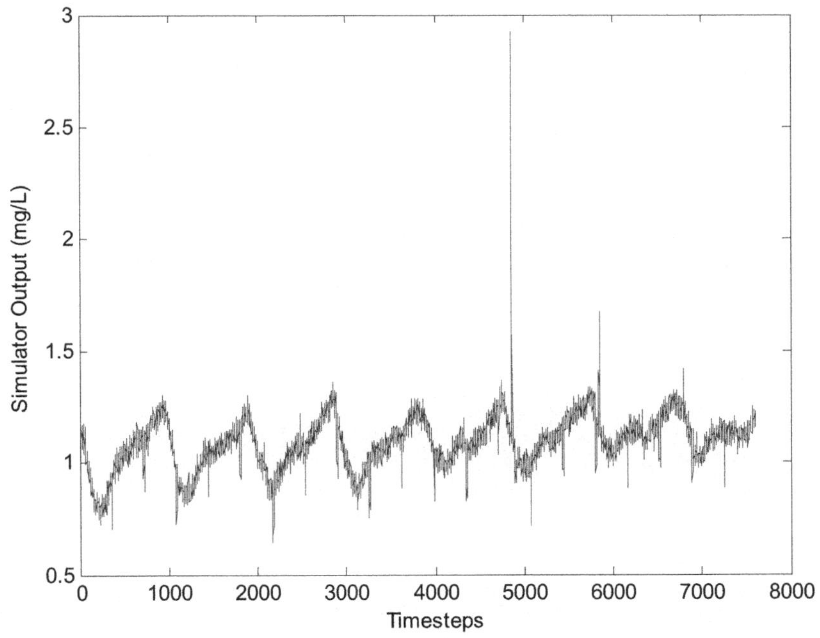

Figure 3.2 Data output from sensor simulator configured with CLX residual chlorine sensor operating parameters. Data is from Run 6, a short (15-min) pulse of sodium fluoroacetate at a NOAEL dose with water quality baseline data included. The x-axis is time and the y-axis is sensor response

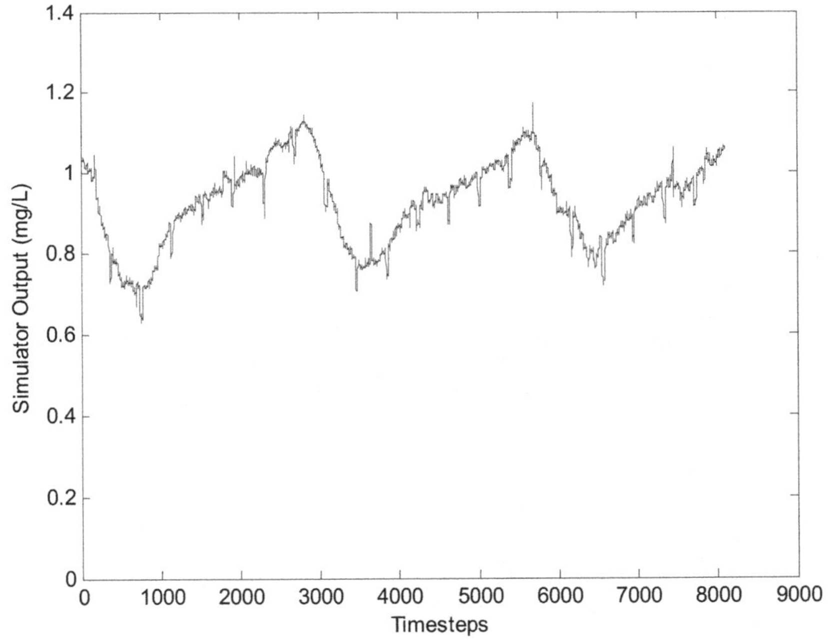

Figure 3.3 Data output from the sensor simulator configured with Endress Hauser residual chlorine sensor operating parameters. Data is from Run 10, a short (15-min) pulse of aldicarb at 4 percent of the NOAEL dose with water quality baseline data included. The x-axis is time and the y-axis is sensor response

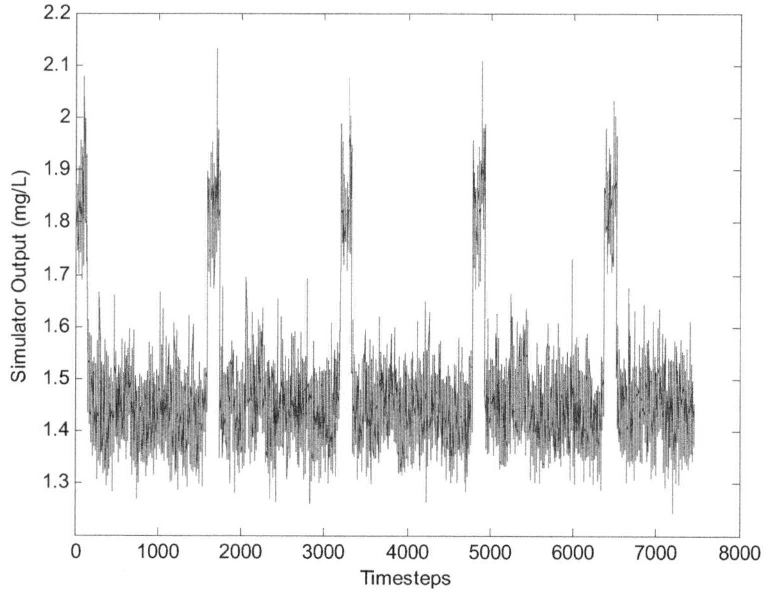

Figure 3.4 Data output from the sensor simulator configured with Hach Total Organic Carbon sensor operating parameters. Data is from Run 9, a long (30-min) pulse of aldicarb at 4 percent of the NOAEL dose with water quality baseline data included. The x-axis is time and the y-axis is sensor response.

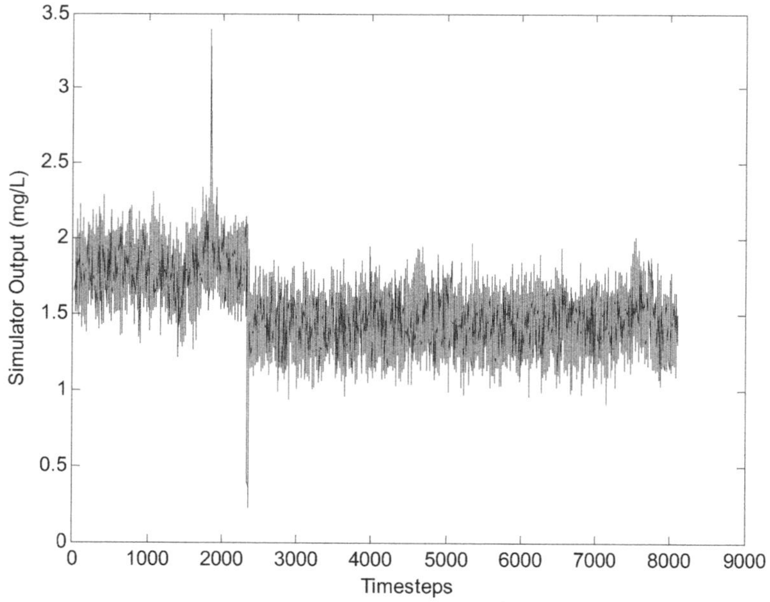

Figure 3.5 Data output from the sensor simulator configured with Specto-Lyser total organic carbon operating parameters. Data is from Run 10, a short (15-min) pulse of aldicarb at 4 percent of the NOAEL dose with water quality baseline data included. The x-axis is time and the y-axis is sensor response.

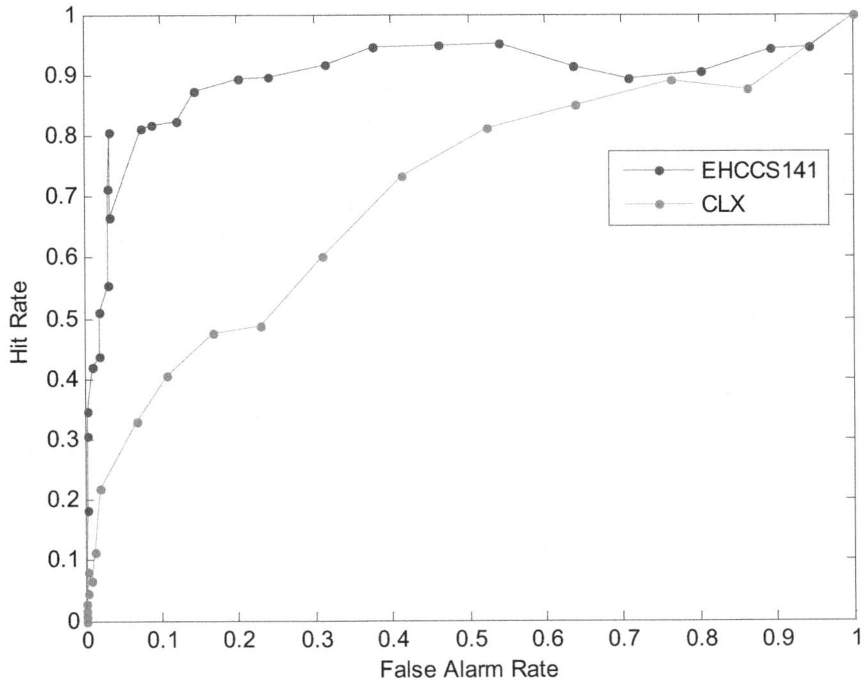

Figure 3.6 ROC curves for chlorine sensors EHCCS141 (ROC Area = 0.89) and CLX (ROC Area = 0.71) using Forward Linear Predictor algorithm on Run 10.

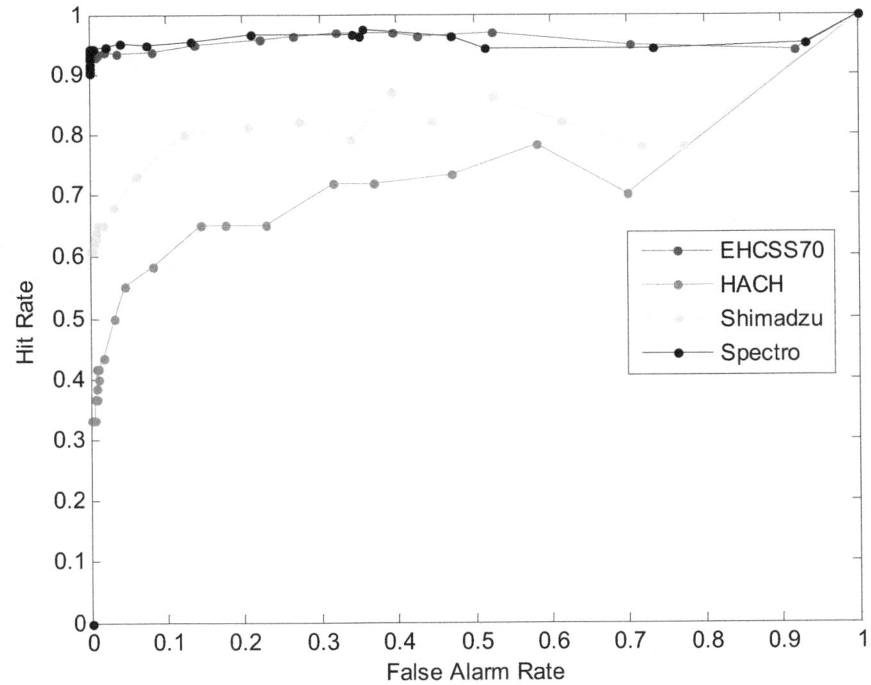

Figure 3.7 ROC curve for TOC sensors EHCSS70 (AUC = 0.96), HACH (AUC = 0.73), Shimadzu (AUC = 0.82) and Spectro (AUC = 0.96) using the Forward Linear Predictor algorithm on Run 2.

Table 3.1
ROC Curve Area Results Using the FLP Algorithm

Run	Description	CLX	EHC	Hach	Spectro	Shimadzu	EHC
1	Aldicarb; baseline on, 30-min pulse	0.96	0.98	0.88	0.97	0.91	0.97
2	Aldicarb, baseline on; 15 min pulse	0.93	0.97	0.73	0.96	0.82	0.96
3	Aldicarb; baseline off; 30-min pulse	0.96	0.96	0.87	0.97	0.92	0.98
4	Aldicarb; baseline off; 15-min pulse	0.93	0.96	0.77	0.96	0.84	0.95
5	NaFluor; Baseline on; 30-min pulse	0.91	0.95	0.87	0.97	0.91	0.97
6	NaFluor; Baseline on; 15-min pulse	0.90	0.96	0.76	0.96	0.85	0.95
7	NaFluor; Baseline off; 30-min pulse	0.94	0.94	0.88	0.97	0.93	0.98
8	NaFluor; Baseline off; 15-min pulse	0.93	0.97	0.82	0.97	0.87	0.96
9	Aldicarb low dose; baseline on; 30-min pulse	0.67	0.83	0.68	0.56	0.70	0.74
10	Aldicarb (low dose); baseline on; 15-min pulse	0.71	0.89	0.63	0.57	0.72	0.73
11	Aldicarb (low dose); baseline off; 30-min pulse	0.92	0.87	0.73	0.58	0.83	0.79
12	Aldicarb low dose; baseline off; 15-min pulse	0.88	0.96	0.75	0.57	0.80	0.77
Runs 1-12 All Avg.	N/A	0.89	0.94	0.78	0.83	0.84	0.90
Runs 1-12 Baseline On Avg.	N/A	0.85	0.93	0.76	0.83	0.82	0.89
14	Aldicarb (very low dose); baseline on; 30-min pulse	0.52	0.56	0.55	0.52	0.52	0.52
15	Aldicarb (very low dose); baseline on; 15-min pulse	0.51	0.53	0.53	0.51	0.55	0.50

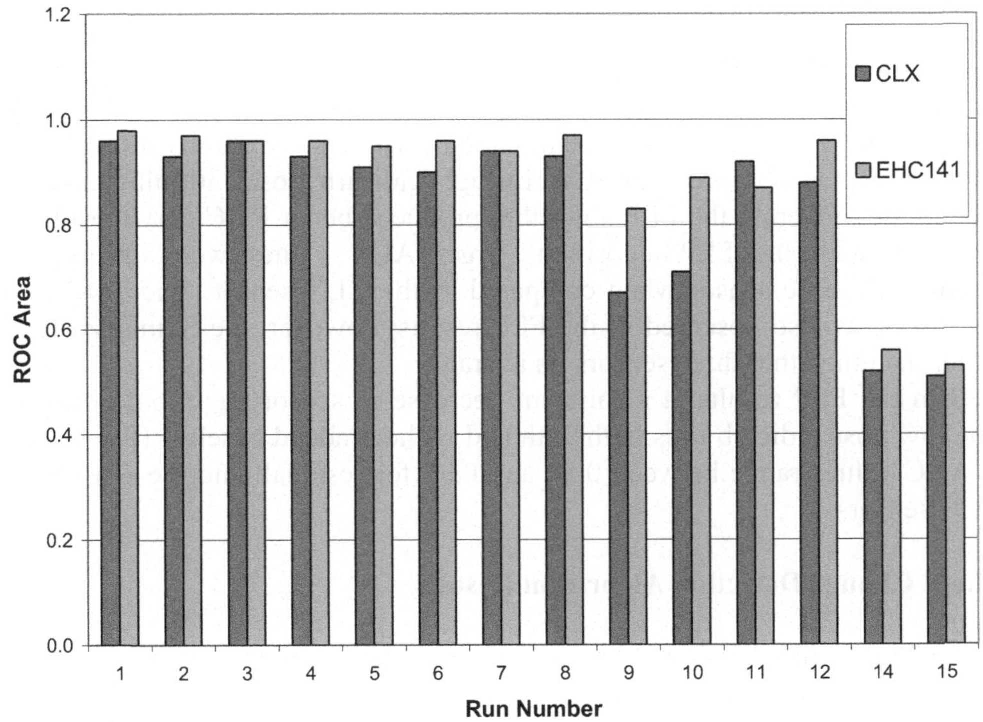

Figure 3.8 ROC curve areas for chlorine sensors using the FLP algorithm.

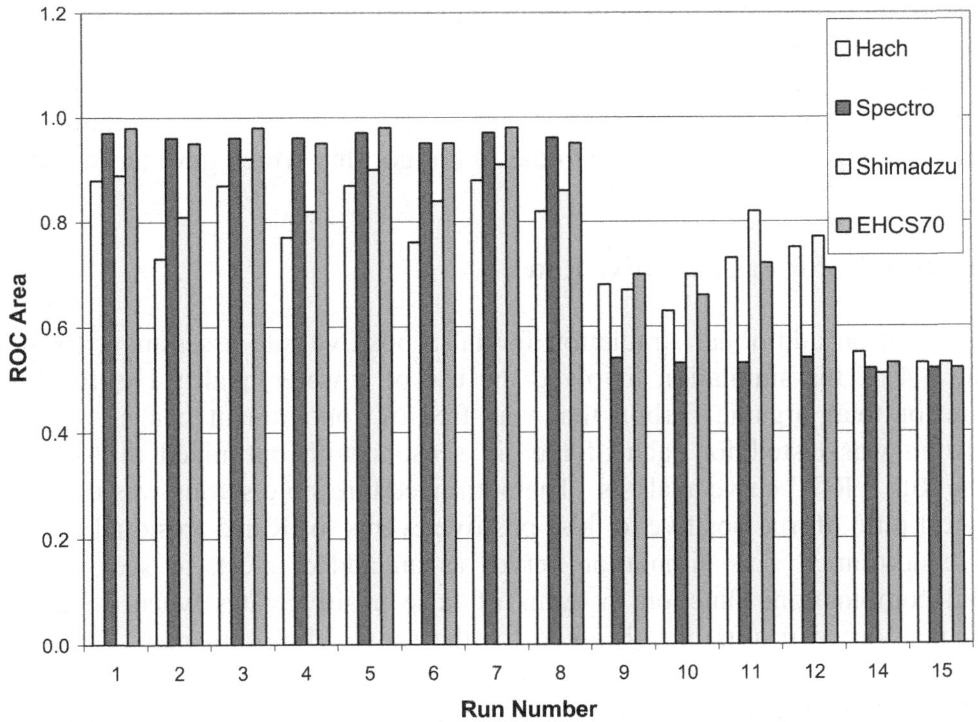

Figure 3.9 ROC curve areas for TOC sensors using the FLP algorithm.

Multivariate Change Detection Algorithm Results

The multivariate (MV) ROC analysis results are shown in Table 3.2 and in Figures 3.10 and 3.11 for the chlorine and TOC sensors, respectively. Typical ROC curves for the Endress Hauser chlorine sensor for two different contamination scenarios are shown in Figure 3.12. A near-perfect ROC curve is observed for Run 1 (high aldicarb dose, 30-min pulse, baseline included) in this figure. Overall, the FLP algorithm produced better ROC curve results than the MV algorithm. As with the FLP algorithm, larger AUC values were observed for the EHCCS141 residual chlorine sensor when compared to the CLX sensor. The TOC results are similar for all sensors, as also observed in the FLP results; however, the Shimadzu ROC areas are slightly higher than the other three sensors on average.

As noted in the FLP results, a significant decrease in sensor-algorithm performance is observed for the low-dose aldicarb runs both with and without added baseline (Runs 9-12). The corresponding AUC values range between 0.53 and 0.97 for residual chlorine sensors and 0.53 and 0.77 for TOC sensors.

Simple Threshold Change Detection Algorithm Results

Results for the simple threshold change detection algorithm are given in Table 3.3 and Figures 3.13 and 3.14 for the chlorine and TOC sensors, respectively. The chlorine sensors performed well with the threshold algorithm at all of the higher contaminant dose levels. In general, the CLX sensor shows higher AUC values with the threshold method than the FLP and MV algorithms, with an overall average of 0.92 and a baseline only average of 0.85. The EHCCS141 sensor had a comparable overall baseline of 0.92 and a baseline only average of 0.85 and was also comparable to the FLP and MV results. The TOC sensors also performed very well with the threshold algorithm. The Hach sensor gave an overall average of 0.88 and a baseline-only average of 0.89. This is much higher than any other sensor/algorithm combination. The closest areas are for the EHCSS70 sensor with the FLP algorithm which gave an overall average of 0.79 and a baseline-only average of 0.80.

Simulation Run Results for False Positive Rates

The results of an additional simulation for false positive rate assessment (Run 13) are shown in Table 3.4. This simulation involved the use of a water quality baseline for either residual chlorine or total organic carbon that was free of contaminant peaks. Each change detection algorithm was tested for each of the six sensors with optimized threshold settings derived from the Run 5 ROC curve analysis. The optimized threshold setting was the setting that yielded the highest ratio of true positive to false positive events for Run 5 (residual chlorine) and Run 9 (TOC) conditions. This simulation serves as an added check to see how well the sensor/algorithm combinations function in terms of false positive rates when no contaminant peaks are present. Varied results are observed for the sensor/algorithm pairs. False positive rates well in excess of 0.9 are observed for the CLX sensor with the FLP and Threshold algorithms. The Endress Hauser chlorine sensor false positive rates fall in a more favorable range (less than 0.004) for both the FLP and the Threshold algorithm. It is apparent that the algorithm chosen in these no-contaminant simulations can have a significant effect on the resulting false positive rates.

Table 3.2
ROC Curve Area Results Using the Multivariate Algorithm

Run	Description	CLX	EHC	Hach	Spectro	Shimadzu	EHC
1	Aldicarb; baseline on, 30 min pulse	0.91	0.95	0.85	0.97	0.89	0.98
2	Aldicarb, baseline on; 15 min pulse	0.91	0.96	0.74	0.96	0.81	0.95
3	Aldicarb; baseline off; 30 min pulse	0.95	0.98	0.85	0.96	0.92	0.98
4	Aldicarb; baseline off; 15 min pulse	0.91	0.96	0.74	0.96	0.82	0.95
5	NaFluor; Baseline on; 30 min pulse	0.88	0.95	0.85	0.97	0.90	0.98
6	NaFluor; Baseline on; 15 min pulse	0.89	0.96	0.73	0.95	0.84	0.95
7	NaFluor; Baseline off; 30 min pulse	0.94	0.98	0.86	0.97	0.91	0.98
8	NaFluor; Baseline off; 15 min pulse	0.91	0.97	0.76	0.96	0.86	0.95
9	Aldicarb (low dose); baseline on; 30 min pulse	0.61	0.78	0.65	0.54	0.67	0.70
10	Aldicarb (low dose); baseline on; 30 min pulse	0.62	0.88	0.62	0.53	0.70	0.66
11	Aldicarb (low dose); baseline off; 30 min pulse	0.86	0.97	0.70	0.53	0.82	0.72
12	Aldicarb low dose; baseline off; 15 min pulse	0.86	0.97	0.72	0.54	0.77	0.71
Runs 1-12 All Avg.	N/A	0.86	0.94	0.76	0.82	0.83	0.88
Runs 1-12 Baseline On Avg.	N/A	0.80	0.91	0.74	0.82	0.80	0.87
14	Aldicarb (very low dose); baseline on; 30 min pulse	0.52	0.47	0.54	0.52	0.51	0.53
15	Aldicarb (very low dose); baseline on; 15 min pulse	0.55	0.55	0.53	0.52	0.53	0.52

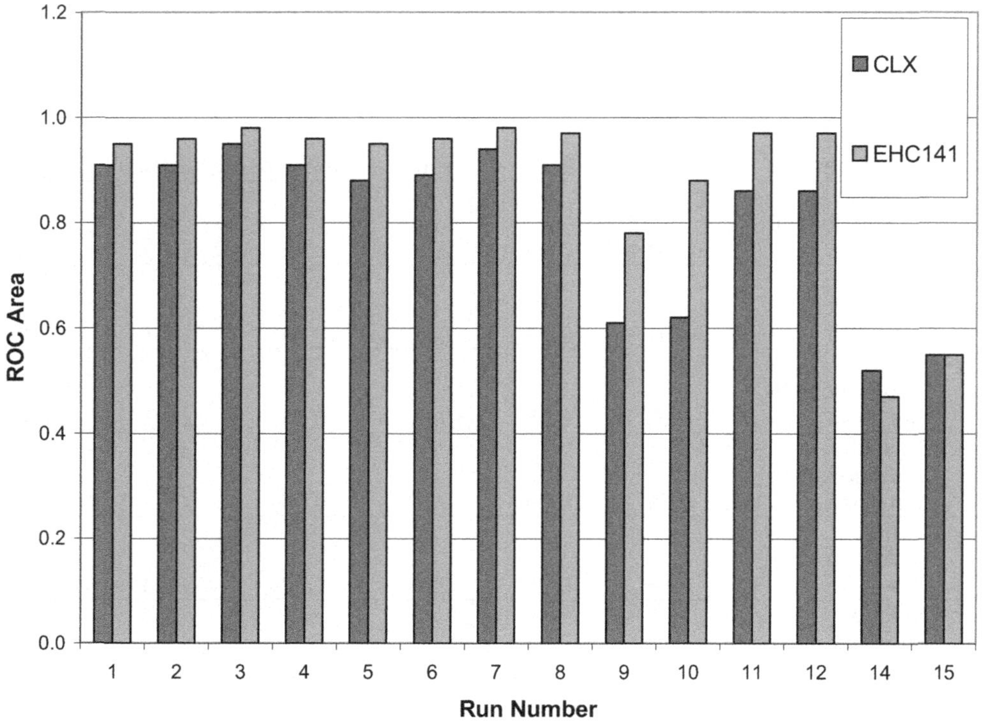

Figure 3.10 ROC curve areas for chlorine sensors using the Multivariate algorithm.

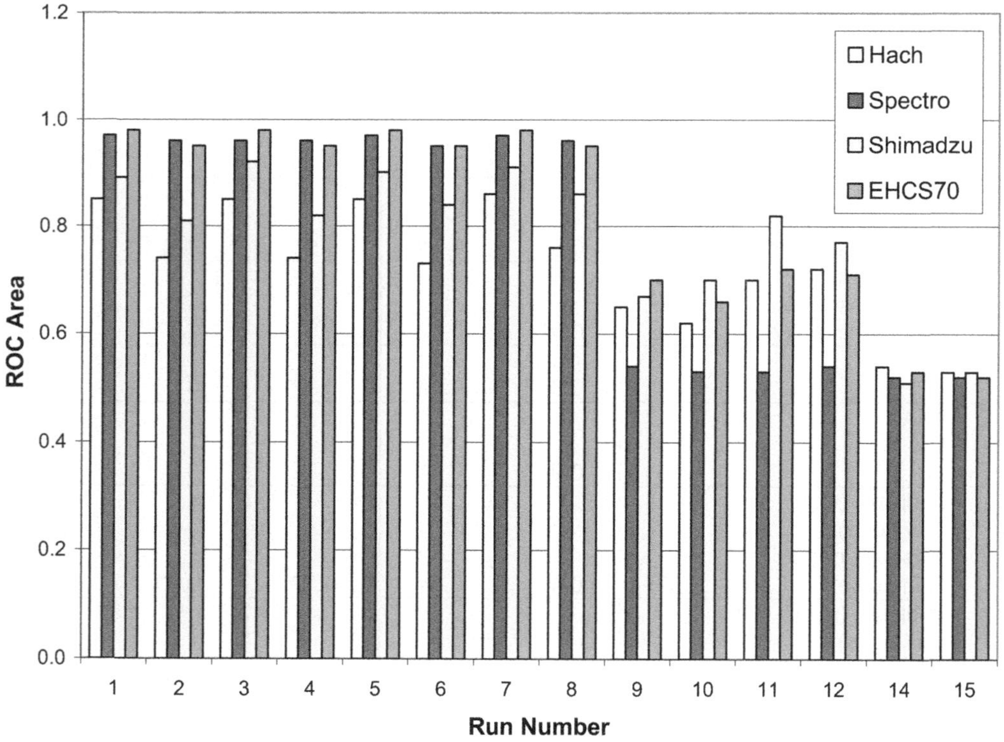

Figure 3.11 ROC curve areas for TOC sensors using the Multivariate algorithm.

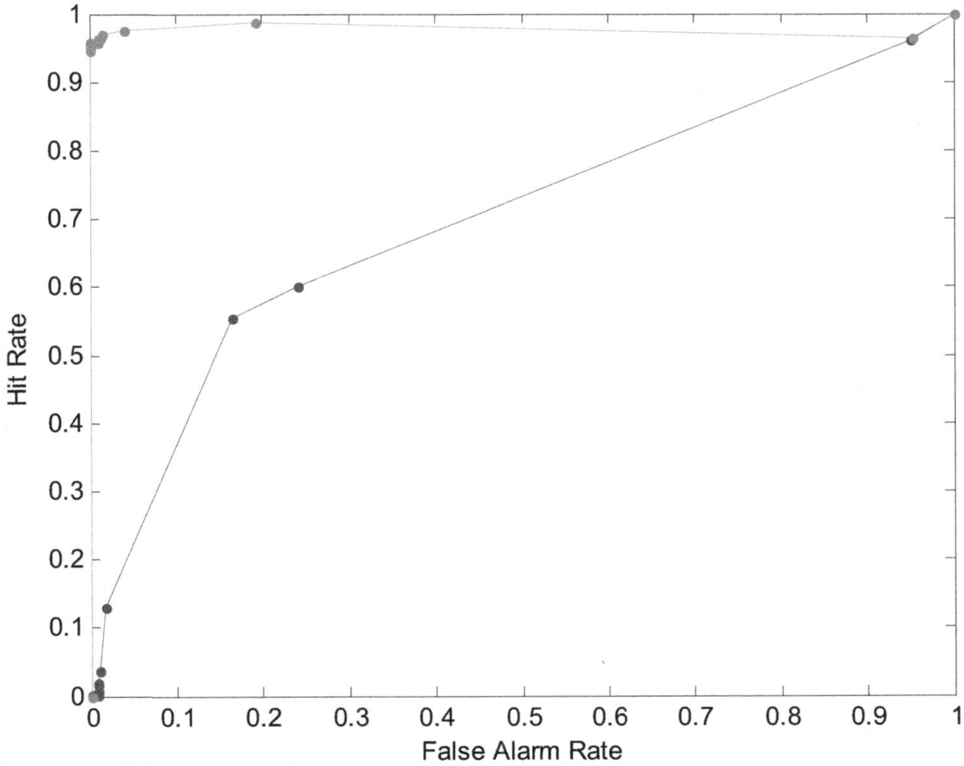

Figure 3.12 ROC curve for EHCSS70 Run 1 (AUC = 0.98) in red and Run 9 (AUC = 0.70) in blue using the Multivariate detection algorithm.

Table 3.3 ROC Curve Area Results Using the Threshold Algorithm

Run #	Description	CLX	EHC	Hach	Spectro	Shimadzu	EHC
1	Aldicarb; baseline on, 30 min pulse	0.99	0.99	1.00	0.98	1.00	0.99
2	Aldicarb, baseline on; 15 min pulse	0.97	0.99	0.92	0.99	0.93	0.99
3	Aldicarb; baseline off; 30 min pulse	1.00	1.00	1.00	0.99	1.00	1.00
4	Aldicarb; baseline off; 15 min pulse	0..99	1.00	0.95	0.98	0.94	0.99
5	NaFluor; Baseline on; 30 min pulse	0.87	0.85	0.99	0.99	1.00	0.99
6	NaFluor; Baseline on; 15 min pulse	0.87	0.87	0.93	0.99	0.94	0.98
7	NaFluor; Baseline off; 30 min pulse	0.98	1.00	1.00	0.99	1.00	1.00
8	NaFluor; Baseline off; 15 min pulse	1.00	1.00	0.98	0.99	0.95	0.99
9	Aldicarb (low dose); baseline on; 30 min pulse	0.70	0.67	0.80	0.63	0.69	0.74
10	Aldicarb (ow dose); baseline on; 30 min pulse	0.68	0.68	0.84	0.62	0.71	0.72
11	Aldicarb (low dose); baseline off; 30 min pulse	0.97	1.00	0.88	0.65	0.78	0.87
12	Aldicarb low dose; baseline off; 15 min pulse	0.96	1.00	0.97	0.65	0.73	0.86
Runs 1-12 All Avg.	N/A	0.92	0.92	0.88	0.87	0.89	0.93
Runs 1-12 Baseline On Avg.	N/A	0.85	0.84	0.97	0.87	0.88	0.90
14	Aldicarb (very low dose); baseline on; 30 min pulse	0.56	0.53	0.41	0.50	0.47	0.51
15	Aldicarb (very low dose); baseline on; 15 min pulse	0.51	0.53	0.48	0.51	0.45	0.51

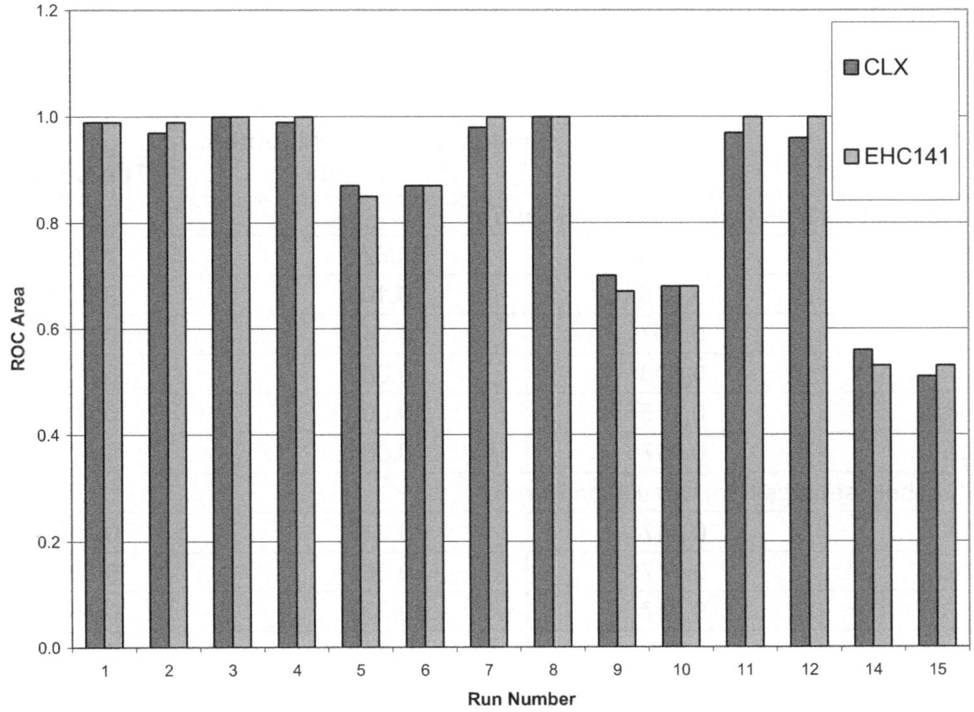

Figure 3.13 ROC curve areas for chlorine sensors using the Threshold algorithm.

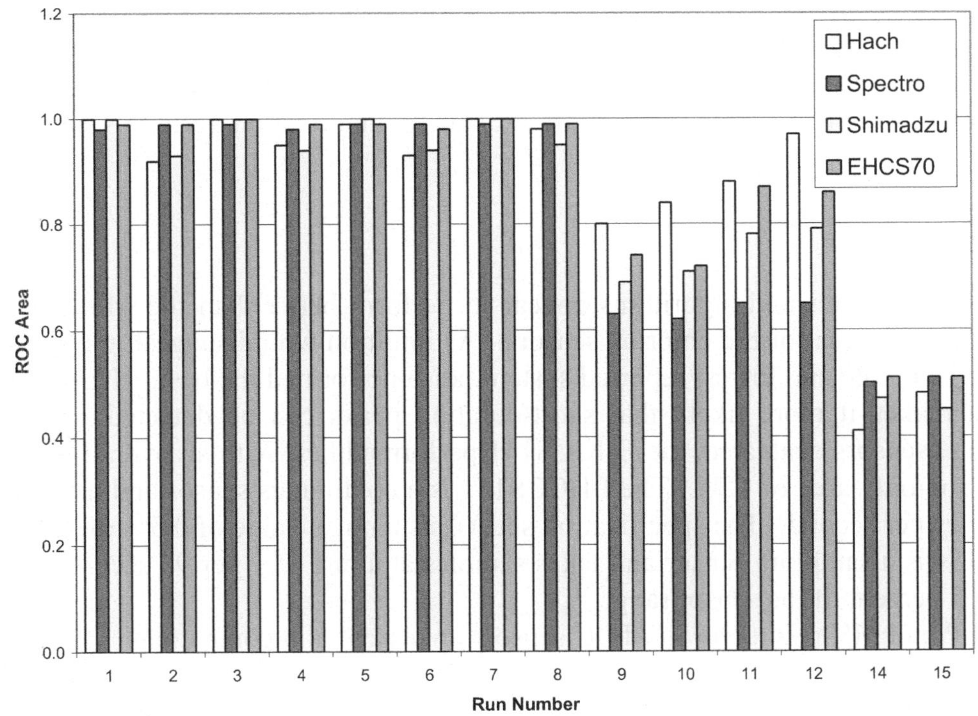

Figure 3.14 ROC curve areas for TOC sensors using the Threshold algorithm.

Table 3.4
False Positive Rates for a Contaminant-free Simulation Run

Sensor	Change Detection Algorithm		
	FLP	Multivariate	Threshold
Optimized threshold settings from Simulation Run 5			
Chlorine-CLX	0.078	0.021	0.164
Chlorine-EHC	0.016	0.104	0.270
TOC-EHC	0.007	0.016	0.008
TOC-Hach	0.178	0.082	0.003
TOC-Shimadzu	0.115	0.000	0.007
TOC-Spectro	0.017	0.001	0.008
Optimized threshold settings from Simulation Run 9			
Chlorine-CLX	0.347	0.362	0.347
Chlorine-EHC	0.177	0.367	0.321
TOC-EHC	0.247	0.532	0.346
TOC-Hach	0.382	0.412	0.276
TOC-Shimadzu	0.390	0.105	0.346
TOC-Spectro	0.385	0.676	0.414

For TOC sensors, false positive rate results are varied, ranging from a high of 0.5 to below 0.001. As noted for the chlorine sensors, the change detection algorithm used in the simulation appears to strongly influence the outcome. These data illustrate that the results of this simulation can be used in parallel with the ROC curve analysis to select the best sensor/algorithm combination within the context of the contamination scenarios chosen in this study.

DISCUSSION

The FLP and threshold algorithms appear to perform better than the MV algorithm in terms of ROC curve areas for the chlorine sensors. As intuition would suggest, the sensors with the fastest sampling rate and lowest reported system noise performed the best. Higher frequency sampling data makes it more likely that short-duration peaks can be detected. For the two residual chlorine sensors evaluated, the EHCCS141 performed best with an overall ROC area of 0.88 and baseline-only area of 0.75. The EHCCS141 is a continuous sensor and has a sampling rate of 2 samples per second. The specification sheet lists a noise value of 0.5% of the measured value. The CLX is a sampling sensor and takes samples at a rate of 1 per 90 seconds and lists a noise value of 5 percent of the sensor range.

The FLP and MV results for the TOC sensors are essentially comparable. The EHCSS70 and the Shimadzu are both very close in terms of ROC curve area statistics. The EHCSS70 captures the shorter pulses better than the Shimadzu since its sampling frequency is higher. The EHCCS70 is a continuous sensor with a sampling rate of 1 sample every 40 seconds and a noise value of 2.5% of the upper range. The Shimadzu is a sampling sensor with a sampling rate of 1 sample every 4 minutes and a noise value of 2% of the full range. In contrast, the Hach sensor is a sampling sensor with a sampling rate of 1 sample every 8 minutes and a noise value of 2% of the upper range. This lower frequency sampling rate is inadequate to capture the relatively short pulse contamination events investigated in this study. The Spectro has a data sampling rate of 1 sample every 30 seconds and a noise value equivalent to 0.3 mg TOC/L. For the shorter contaminant pulse length the Spectro outperforms the Shimadzu. Overall the EHCCS141 and the EHCCSS70 provide the best performance in terms of ROC area results. The threshold algorithm performed well on the TOC sensors. The HACH sensor provided the highest ROC areas on average with this particular algorithm. For many of the runs the contaminant pulse is clearly observable above the noise and is easily detected with the threshold algorithm.

The dose levels we investigated in this study ranged from relatively high (but still non-toxic from a dose-response perspective for the two chemicals studied) to very low. The results indicate that most of the sensor/algorithm combinations can perform adequately at the NOAEL dose level for the contaminants chosen for study. Performance for all sensor/algorithm combinations dropped off considerably when the aldicarb dose level was decreased 25-fold. Reduced concentration levels can lead to a either a diminished ability to reliably detect an event or an increased likelihood of false positive events. Further study is required to more fully characterize the performance of sensors and change detection systems at these lower concentration levels.

The false alarm evaluation (Run 13) reveals quite high false alarm rates for many sensor/algorithm combinations. Many false alarm rates are observed to be in excess of 1 percent for the two baseline conditions evaluated. For a sensor system collecting samples at 30-second intervals, this translates to about 28 false alarms per 24-hour period for a single sensor system. This number of false alarms is unacceptable from a utility operational perspective. False alarms, until formally dismissed as such, require operator response and further investigation which can tie up limited utility personnel resources and potentially erode operator as well as public confidence in the overall utility operation. One promising approach for reducing the number of false alarms in the event detection algorithms is to incorporate information from multiple sensors and from multiple time steps to provide additional confidence with regard to the occurrence of a true event (McKenna et al. 2007).

However, as a further consideration with respect to false alarm rates, this study did not investigate the positive influences of 1) multiple orthogonal sensors at the same location, 2) similar sensors at multiple locations, or combinations of both 1 and 2. Both could arguably improve detection statistics by reason of anticipated spatial and temporal data correlation. If the presence of a contaminant can be detected with a TOC sensor and a chlorine sensor, then the false alarm rate could potentially be reduced. In other words, for a contamination event to be noted, sensor response would be expected on both types of sensors at a particular location. Similarly, if a contaminant is detected at point A within a distribution system, then it can be reasoned that it should also be seen at a downstream location consistent with the anticipated flow behavior of the system under assumed water consumption rates. Failure to detect the event at a downstream location may indicate that the initial detect could be ruled a false positive. Hydraulic modeling software can be used to aid in the correlation of contaminant peaks at multiple locations throughout a distribution system and is the topic of ongoing research efforts (Murray, Janke and Uber 2004; EPA 2005b). In this study we have not quantitatively examined either of these detection refinements; however the simulation tools developed here could be expanded to evaluate the performance of multiple sensor types at the same location as well as similar sensors distributed throughout a network.

CHAPTER 4
SUMMARY AND CONCLUSIONS

In this study a software tool has been developed that enables candidate sensors for intended use in water distribution contamination warning systems to be evaluated using simulation techniques. The tool is intended to be used during the design of a contaminant warning system within the distribution system of a water utility. The tool incorporates sensor performance specifications that are either vendor-provided or obtained through third-party testing, and combines these with user-defined contamination scenarios within a user-specified pipe network in order to assess overall contamination event detection performance and optimize CWS design.

Simulation testing of a selected group of commercially available sensors in combination with several event detection algorithms showed increasingly degraded system performance with decreasing contaminant concentration levels. At the highest dose level examined in this investigation, performance of all sensor systems was quite good. Contaminant dose levels selected for study used a conservative no observable adverse effect level for a carbamate pesticide as the baseline dose and also included two additional dose levels approximately two and three orders of magnitude below the no observable adverse effect level. The good results observed for the highest dose case suggest that current sensor systems can perform adequately in combination with change detection software to detect contamination events for this class of moderately toxic industrial chemicals. At intermediate dose levels, system performance is observed to range from poor to fair with an increasing incidence of false positives anticipated in actual use. At the lowest dose level examined—in the range of micrograms of toxin per liter of water—all sensor/change detection combinations evaluated in this study provided essentially no value as reliable detection devices in a CWS. In these very low dose cases, the performance analysis data shows that the probability of a false positive is essentially equal to the probability of a true positive. Simulations were carried out with and without a water quality baseline for the sensor being evaluated. Results for the simulations at the baseline dose levels indicate that expected sensor performance is not significantly degraded by the inclusion of a water quality baseline.

Some performance differences among sensor systems were observed and the most influential operational parameter of the sensor appears to be the sensor's sampling interval. Those sensors with a higher frequency sampling rate performed better than those with lower sampling rates. Sensors with a lower sampling rate can entirely miss a contamination pulse if it is of short duration. The more advanced statistical change detection algorithms, specifically the FLP method, generally perform better than the threshold technique for residual chlorine sensors. No significant performance improvement was observed with the advanced detection methods over the threshold methods for the total organic carbon sensors for the contamination scenarios investigated in this study. It is important to note that the statistically based methods are designed for application in CWS designs involving multiple different sensor types at multiple locations and that they are expected to perform better than simple threshold methods under those conditions. These more complex, multi-sensor and multi-location configurations were not investigated in this study.

This study illustrates the usefulness of simulation tools that can be used during the design of a contaminant warning system for a water utility. Using such tools, inter-comparisons of candidate sensor performance can be made under a variety of realistic contamination scenarios that can enable utility operators to fully optimize the design prior to CWS system acquisition and deployment. Further software development could be carried out to better integrate the various components of the simulation package. At present, the output from the sensor simulator is imported into a change detection module and responses are compiled with some degree of operator intervention required. Furthermore, the desired contamination scenarios are manually configured and integrated into the sensor simulator. A more fully integrated system with an advanced graphical user interface would provide a user-friendly tool for broader water utility use during CWS design.

The work reported here was carried out using single sensor systems at one location within a distribution system. A more effective tool could be developed that would enable performance assessment of multiple sensors at multiple locations. The basic principles for how to conduct the simulation outlined in this study would not change; however, the simulations would have to be expanded to cover multiple locations. A considerable degree of automation of the simulation process would have to be done since manual processing of data streams from multiple locations and multiple sensors would be very labor intensive and susceptible to data handling errors.

APPENDIX A
TECHNICAL DESCRIPTIONS OF CHANGE DETECTION ALGORITHMS

Multivariate Distance

The multivariate distance (MV) algorithm considers changes in water quality based on the difference between successive water quality measurements in multivariate space. The distance measure used to distinguish between normal and anomalous water quality is based on the Euclidean distance from the normalized water quality measurement to the mean water quality. Normalized and mean water qualities are a function of the size of the moving window. A constant distance threshold, d_c, is applied to the distance measure to determine if the water quality measurement is anomalous. Distances above the threshold are categorized as anomalous, while distances below the threshold are considered normal.

The distance measure (d) used in the MV algorithm is the difference between the Euclidean distances of successive points to the mean in n dimensional space.

$$d = \sqrt{\sum_{j=1}^{n} \left| Z(t)_j \right|^2} - \sqrt{\sum_{j=1}^{n} \left| Z(t-1)_j \right|^2} \qquad (1)$$

where n is the number of water quality parameters, $Z(t)_j$ is the jth water quality parameter in the current time step, and $Z(t-1)_j$ is the jth water quality parameter in the previous time step. The measured multivariate distance is then compared to d_c to determine if it is an anomalous change in water quality. The MV algorithm distinguishes water quality anomalies when the distance to the mean is significantly different from the distance to the mean of the previous time step.

Forward Linear Predictor

A forward linear predictor (FLP) estimates the current value of a time series based on past samples. In its most general form, this approach is also known as the autoregressive moving average predictor (Bras and Rodriguez-Iturbe, 1993) and the most common representation is

$$Z^*(t) = -\sum_{i=1}^{P} a_i Z(P-i), \qquad (2)$$

where $Z^*(t)$ is the predicted signal value, $Z(p-i)$, are previous observed values, a_i the predictor coefficients, and P the order of the prediction filter polynomial. P is also the size of the standardization window as described previously. The error generated by this estimate is

$$e(t) = Z(t) - Z^*(t), \qquad (3)$$

where $Z^*(t)$ is the true signal value. Several methods are available to solve for the values of a_i. In our implementation the autocorrelation method of autoregressive (AR) modeling is used to find the predictor coefficients. In this method we compute the least squares solution to

$$Za \approx b \tag{4}$$

where

$$Z = \begin{bmatrix} Z(1) & 0 & \cdots & 0 \\ Z(2) & Z(1) & \ddots & \vdots \\ \vdots & Z(2) & \ddots & 0 \\ Z(m) & \vdots & \ddots & Z(1) \\ 0 & Z(m) & \ddots & Z(2) \\ \vdots & \ddots & \ddots & \vdots \\ 0 & \ldots & 0 & Z(m) \end{bmatrix}, \quad a = \begin{bmatrix} 1 \\ a(2) \\ \vdots \\ a(p+1) \end{bmatrix}, \quad b = \begin{bmatrix} 1 \\ 0 \\ \vdots \\ 0 \end{bmatrix}$$

and m is the length of Z and is equal to $P+1$. The linear least squares problem is generally solved by applying

$$Z^T Za = Z^T b \tag{5}$$

where Z^T is the transpose of Z. This leads to the Yule-Walker equations

$$\begin{bmatrix} r(1) & r(2) & \cdots & r(p) \\ r(2) & r(2) & \ddots & \vdots \\ \vdots & \ddots & \ddots & r(2) \\ r(p) & \cdots & r(2) & r(1) \end{bmatrix} \begin{bmatrix} a(2) \\ a(3) \\ \vdots \\ a(p+1) \end{bmatrix} = \begin{bmatrix} -r(2) \\ -r(3) \\ \vdots \\ -r(p+1) \end{bmatrix} \tag{6}$$

where r is an autocorrelation estimate for Z computed using

$$r = \frac{1}{m} IFFT[|FFT[Z]|^2] \tag{7}$$

where FFT and $IFFT$ are the forward and inverse discrete Fourier transforms, respectively. Since equation 6 is a Toeplitz matrix, the use of Levinson-Durbin (LD) recursion provides the most efficient solution for a. An ancillary output of the LD recursion is the LD prediction error associated with the uncertainty of the inversion. The LD prediction error is also referred to as the variance, or power, of the prediction error. Once a has been found it is inserted back into equation 2 and the current value of the signal is estimated. The resulting error from equation 3 is compared to the critical threshold which determines if an anomaly has occurred.

APPENDIX B
GRAPHS OF ALL SIMULATIONS

Run Descriptions

(See Table 8 in Main Report)

- Run 1
  - Baseline On, 30 min. pulse, Aldicarb
- Run 2
  - Baseline On, 15 min. pulse, Aldicarb
- Run 3
  - Baseline Off, 30 min. pulse, Aldicarb
- Run4
  - Baseline Off, 15 min. pulse, Aldicarb
- Run 5
  - Baseline On, 30 min. pulse, Nafluoroacetate
- Run 6
  - Baseline On, 15 min. pulse, Nafluoroacetate
- Run 7
  - Baseline Off, 30 min. pulse, Nafluoroacetate
- Run 8
  - Baseline Off, 15 min. pulse, Nafluoroacetate
- Run 9
  - Baseline On, 30 min. pulse, Low Aldicarb
- Run 10
  - Baseline On, 15 min. pulse, Low Aldicarb
- Run 11
  - Baseline Off, 30 min. pulse, Low Aldicarb
- Run 12
  - Baseline Off, 15 min. pulse, Low Aldicarb
- Run 13
  - Baseline On, No Contaminant
- Run 14
  - Baseline On, 30 min. pulse, Low2 Aldicarb
- Run 15
  - Baseline On, 15 min. pulse, Low2 Aldicarb

Sensor Key

Abbreviation	Sensor
CLX	HF Scientific Model CLX (Chlorine)
EHCCS141	Endress Hauser Model CCS141 (Chlorine)
EHCSS70	Endress Hauser Model CSS70 (Total Organic Carbon)
HACH	Hach Model Astro TOC (Total Organic Carbon)
Shimadzu	Shimadzu Model TOCN-4110 (Total Organic Carbon)
Spectro	S:Scan Model Spectro:Lyser TOC

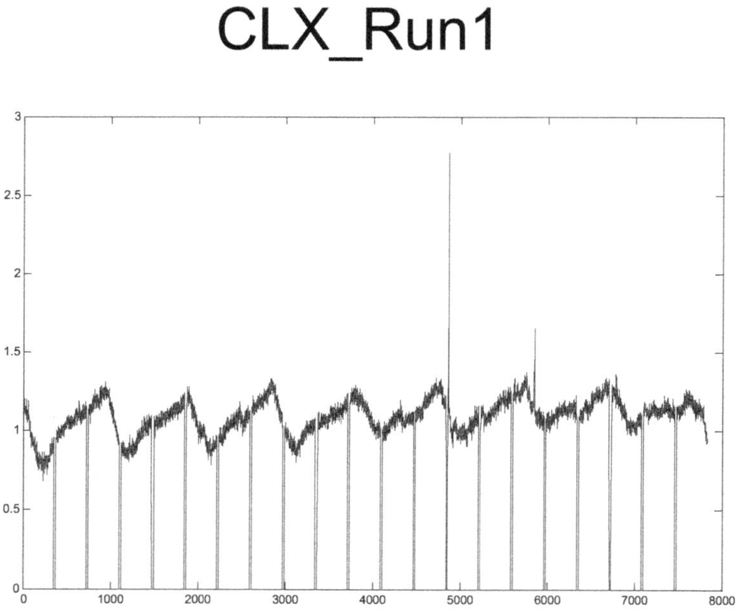

Figure B-1 Baseline On, 30 min. pulse, Aldicarb

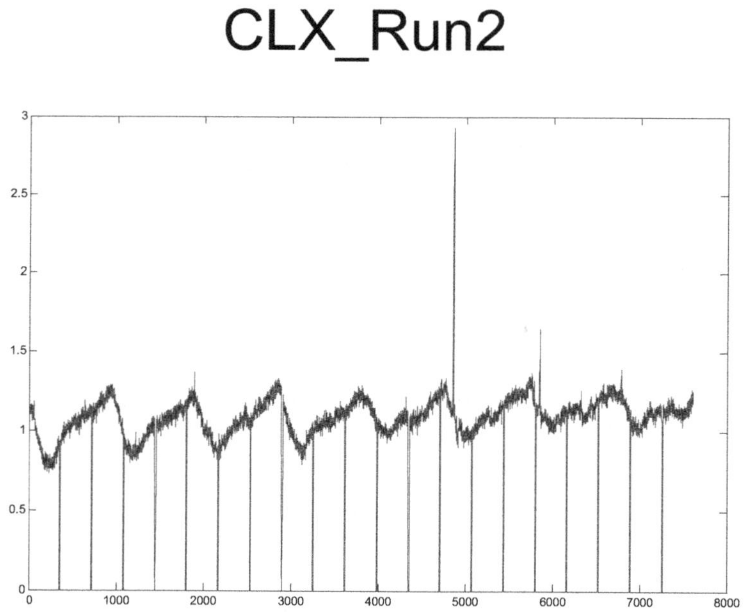

Figure B-2 Baseline On, 15 min. pulse, Aldicarb

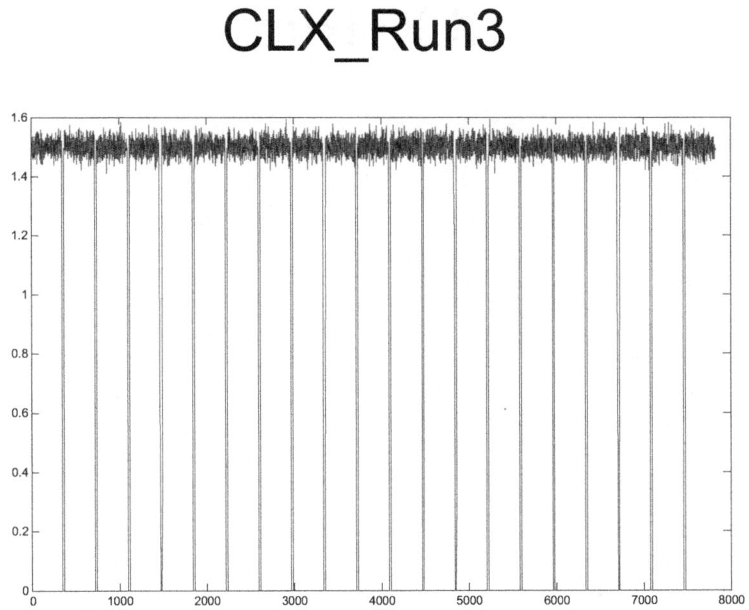

Figure B-3 Baseline Off, 30 min. pulse, Aldicarb

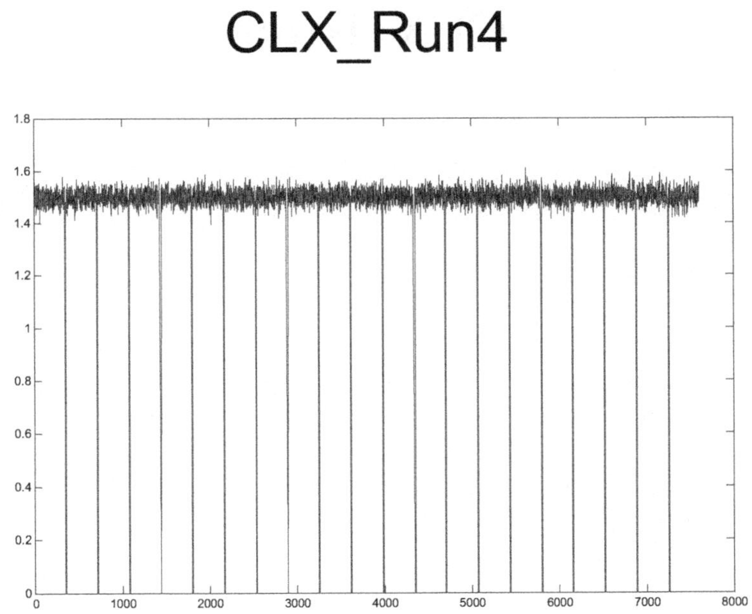

Figure B-4 Baseline Off, 15 min. pulse, Aldicarb

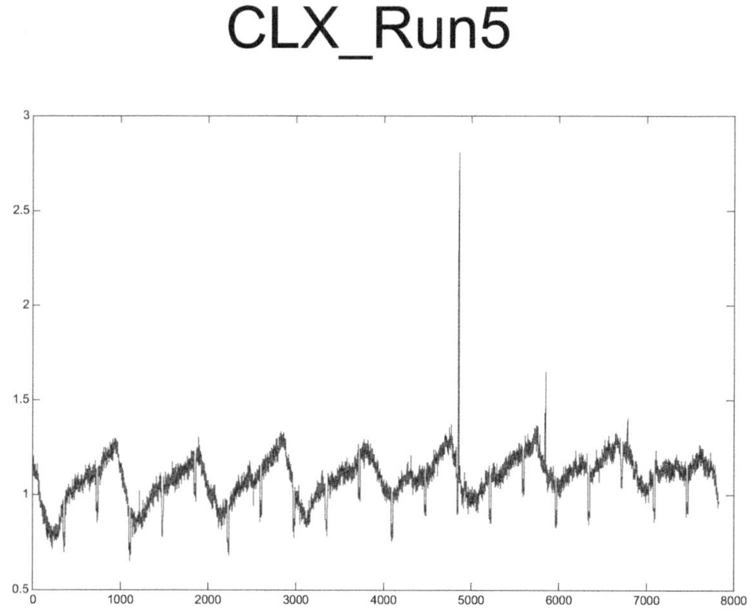

Figure B-5 Baseline On, 30 min. pulse, Sodium Fluoroacetate

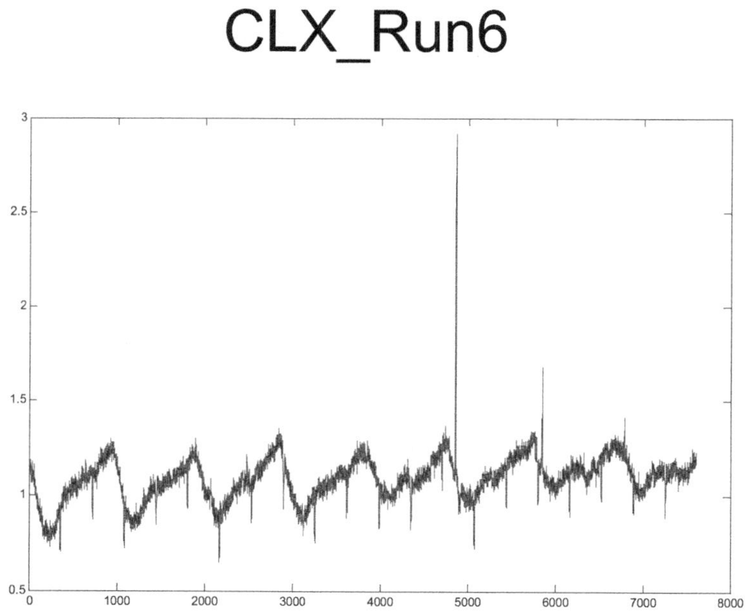

Figure B-6 Baseline On, 15 min. pulse, Sodium Fluoroacetate

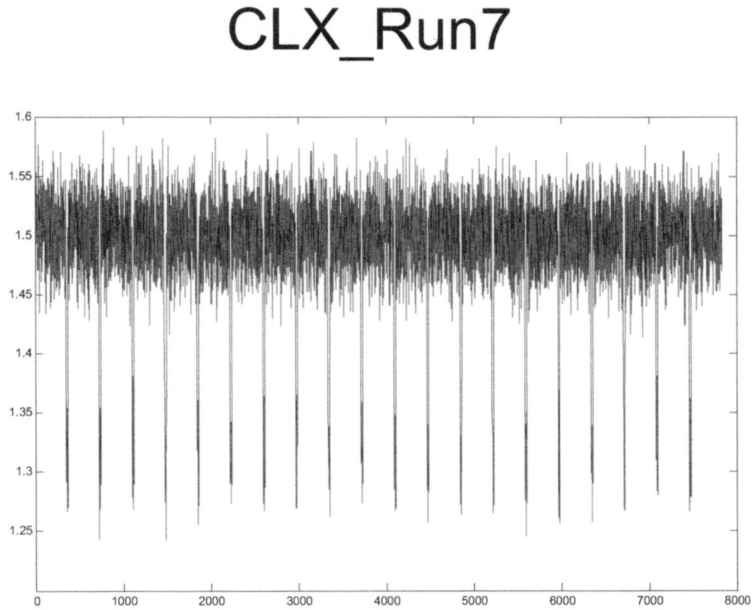

Figure B-7 Baseline Off, 30 min. pulse, Sodium Fluoroacetate

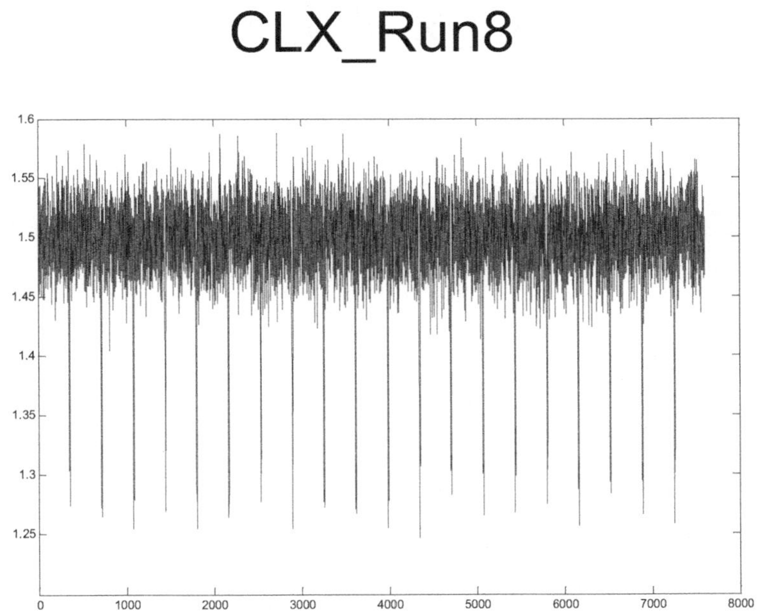

Figure B-8 Baseline Off, 15 min. pulse, Sodium Fluoroacetate

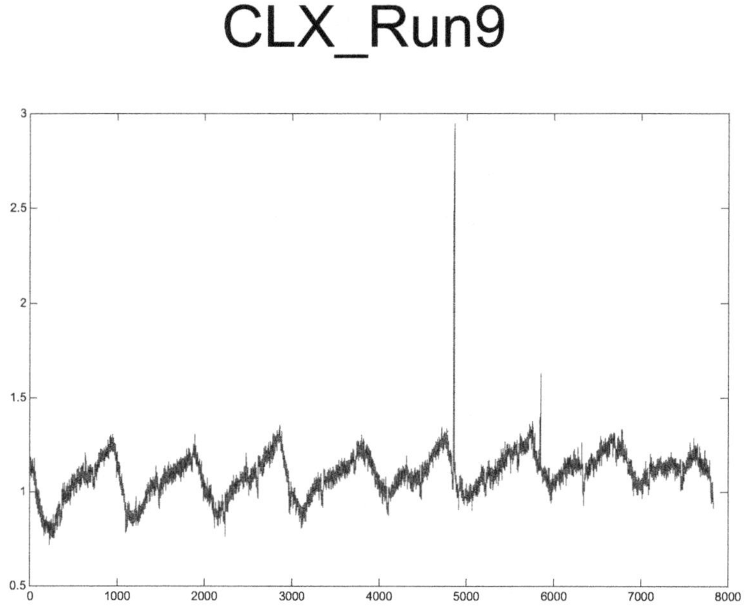

Figure B-9 Baseline On, 30 min. pulse, Low Aldicarb

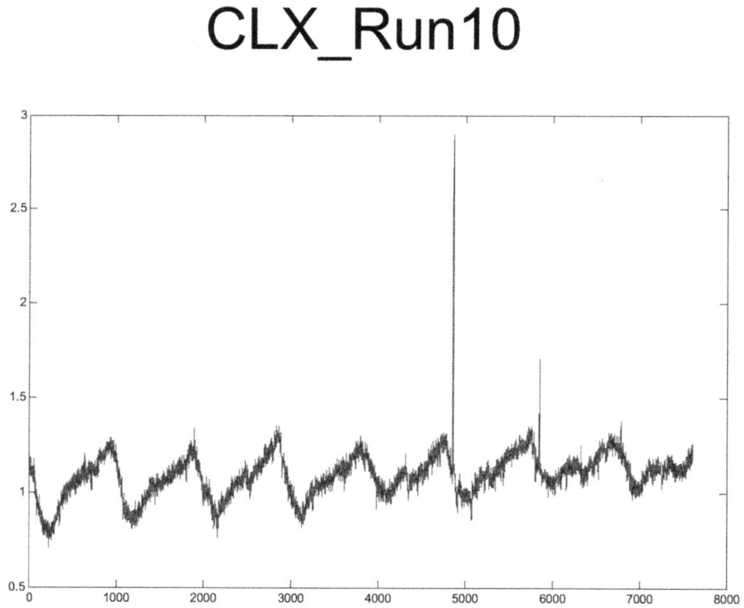

Figure B-10 Baseline On, 15 min. pulse, Low Aldicarb

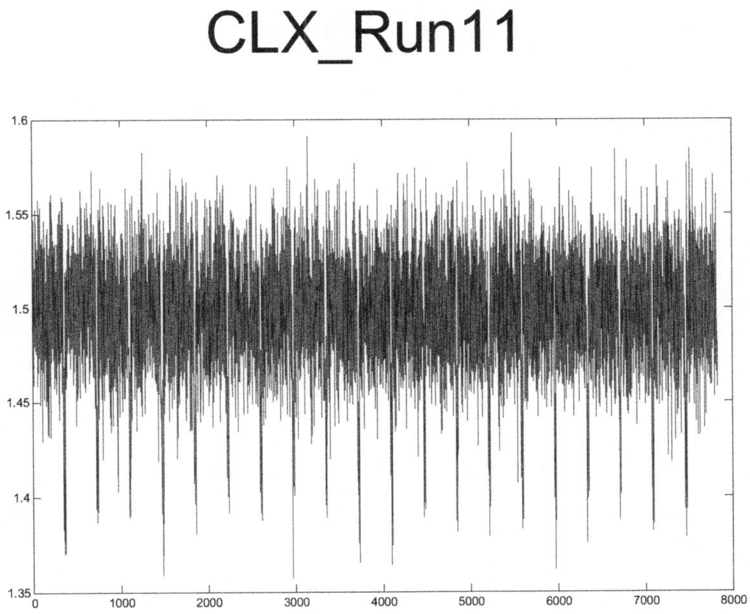

Figure B-11 Baseline Off, 30 min. pulse, Low Aldicarb

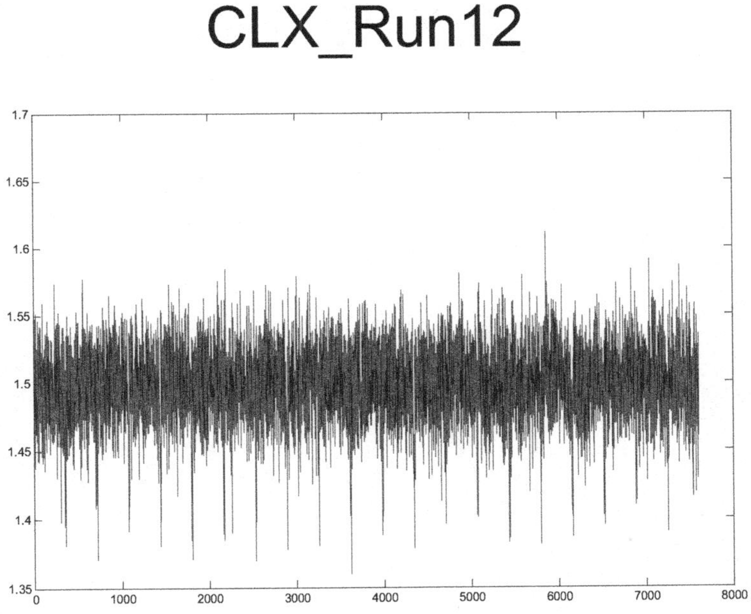

Figure B-12 Baseline Off, 15 min. pulse, Low Aldicarb

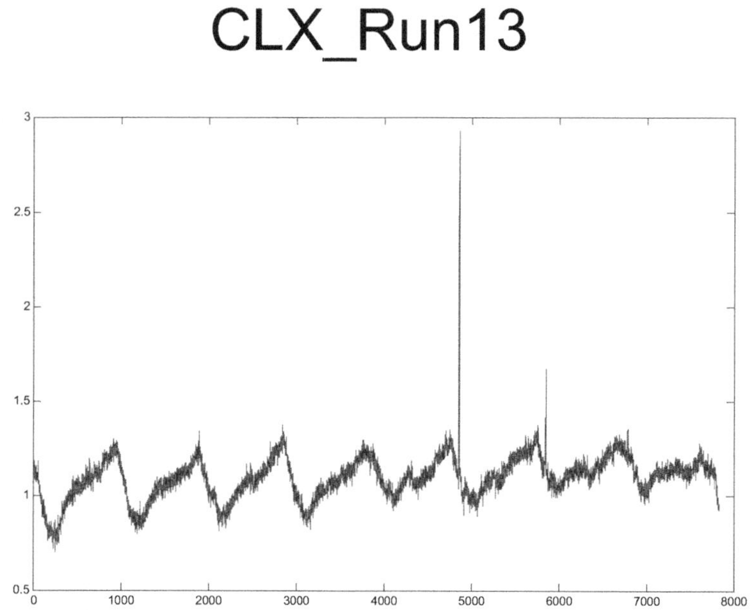

Figure B-13 Baseline On, No contaminant pulse.

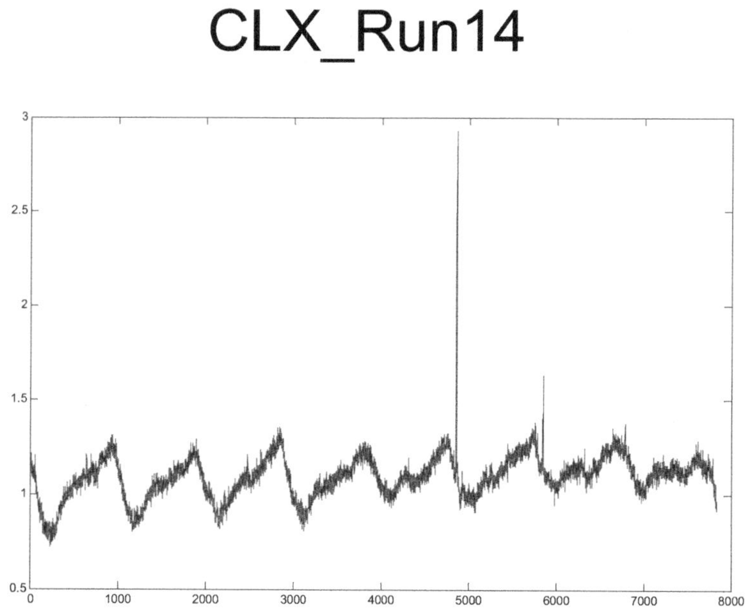

Figure B-14 Baseline On, 30 min. pulse, Low2 Aldicarb

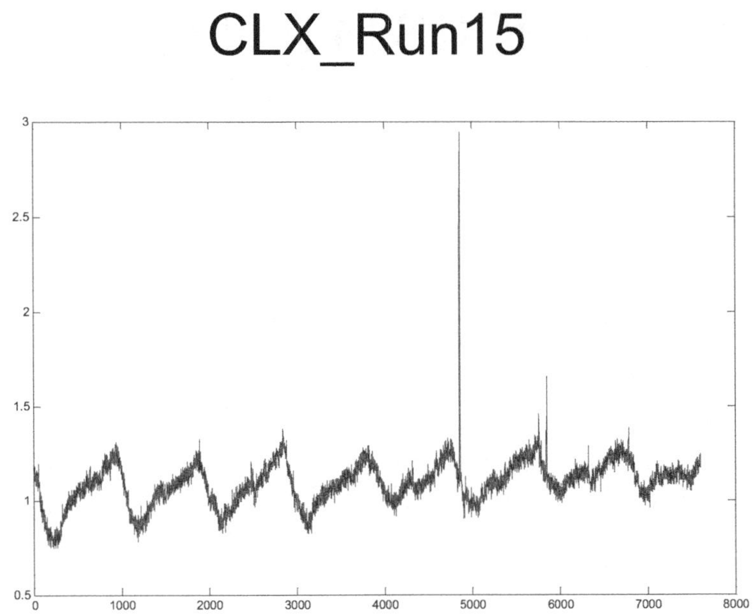

Figure B-15 Baseline On, 15 min. pulse, Low2 Aldicarb

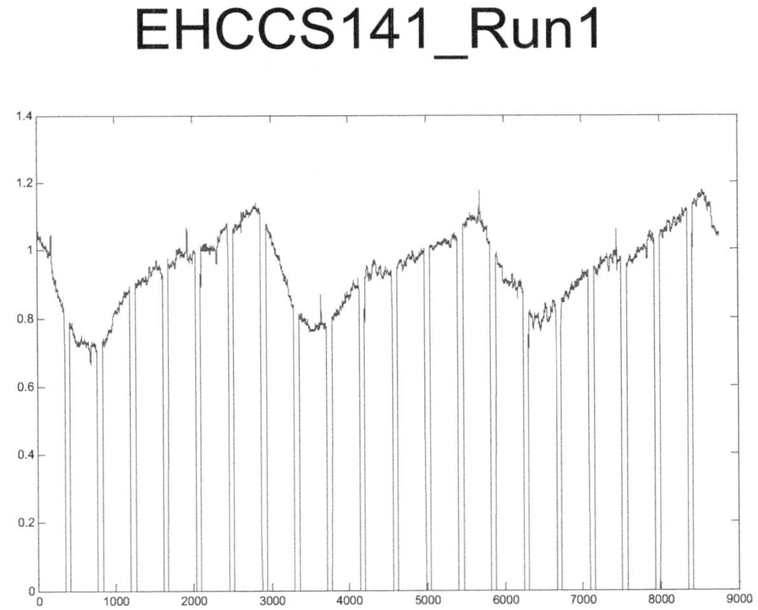

Figure B-16 Baseline On, 30 min. pulse, Aldicarb

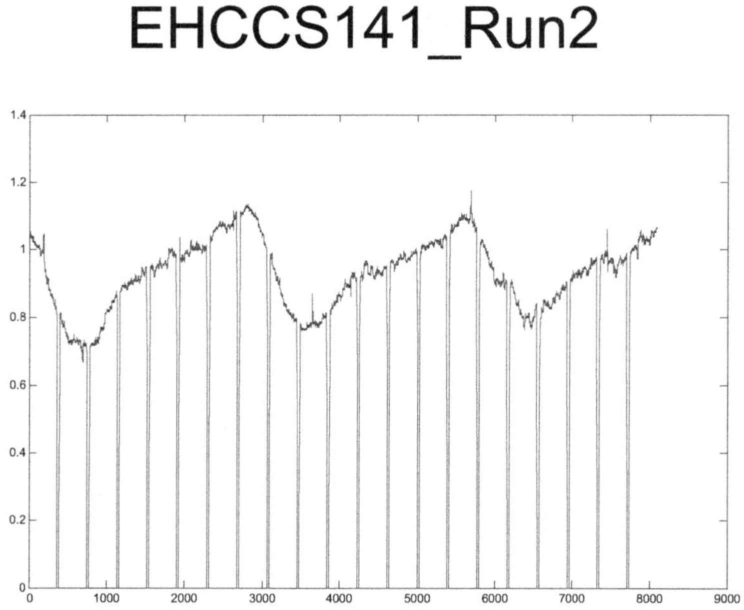

Figure B-17 Baseline On, 15 min. pulse, Aldicarb

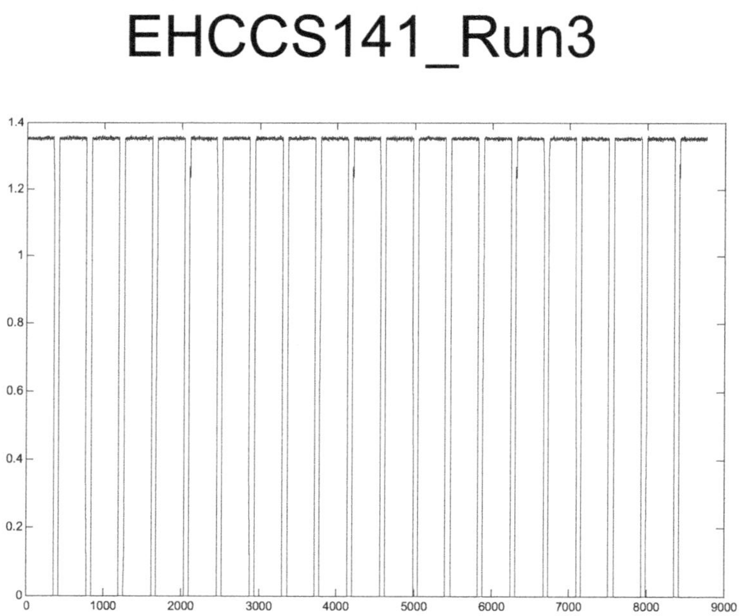

Figure B-18 Baseline Off, 30 min. pulse, Aldicarb

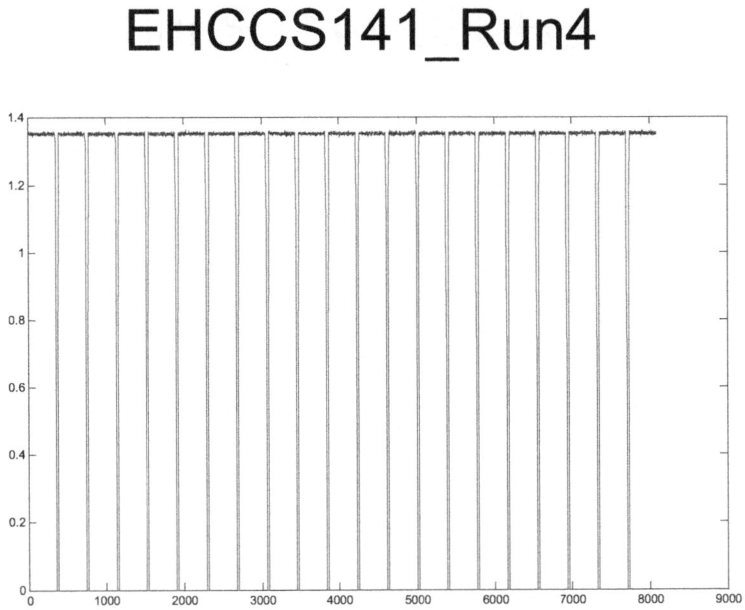

Figure B-19 Baseline Off, 15 min. pulse, Aldicarb

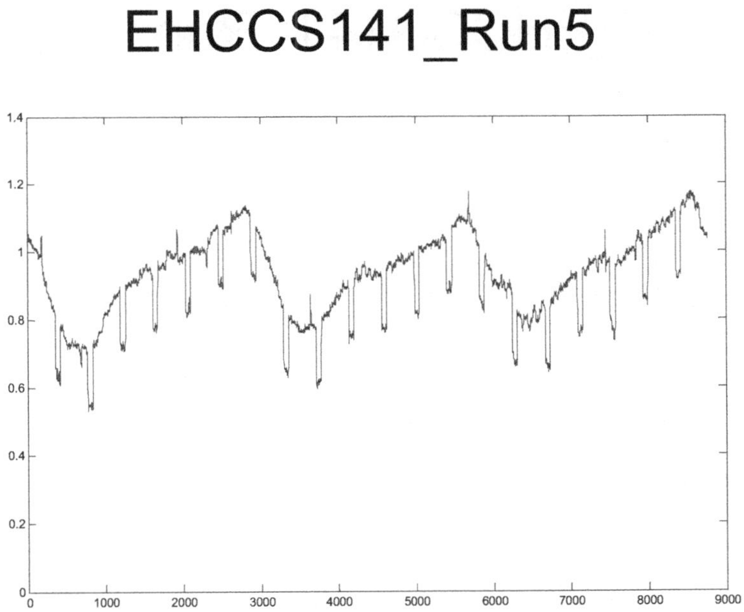

Figure B-20 Baseline On, 30 min. pulse, Sodium Fluoroacetate

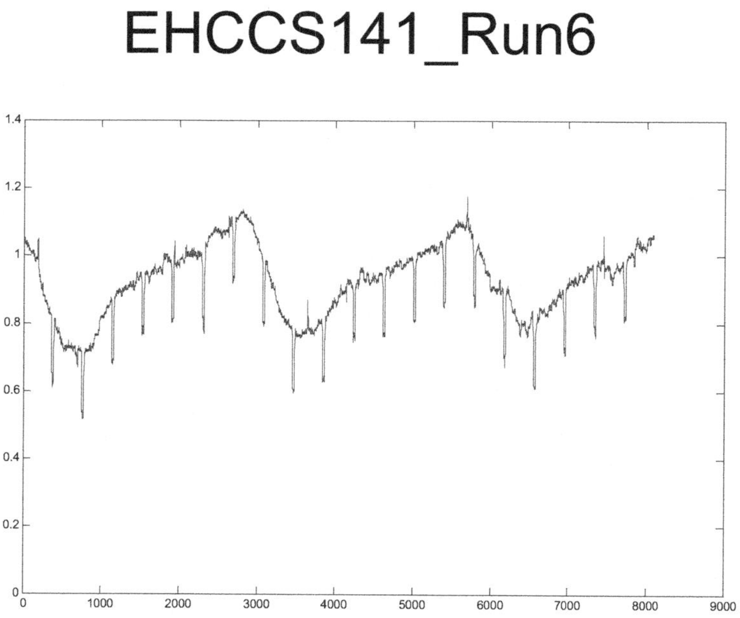

Figure B-21 Baseline On, 15 min. pulse, Sodium Fluoroacetate

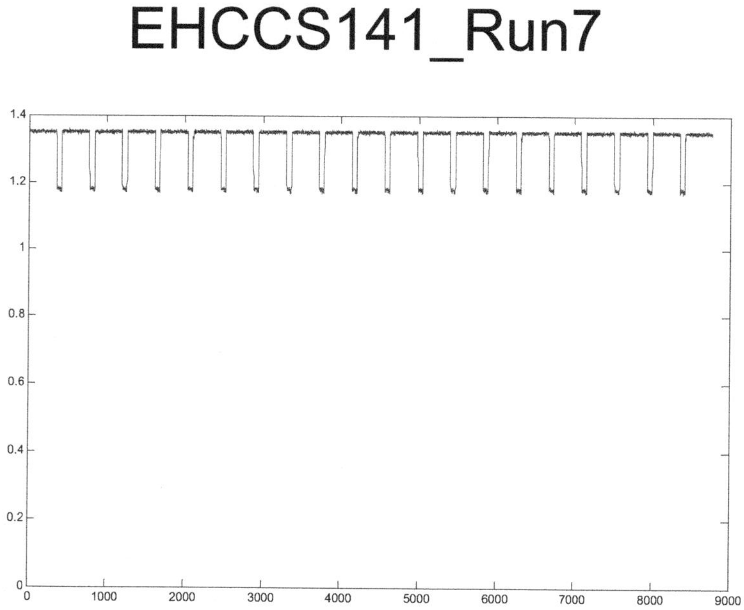

Figure B-22 Baseline Off, 30 min. pulse, Sodium Fluoroacetate

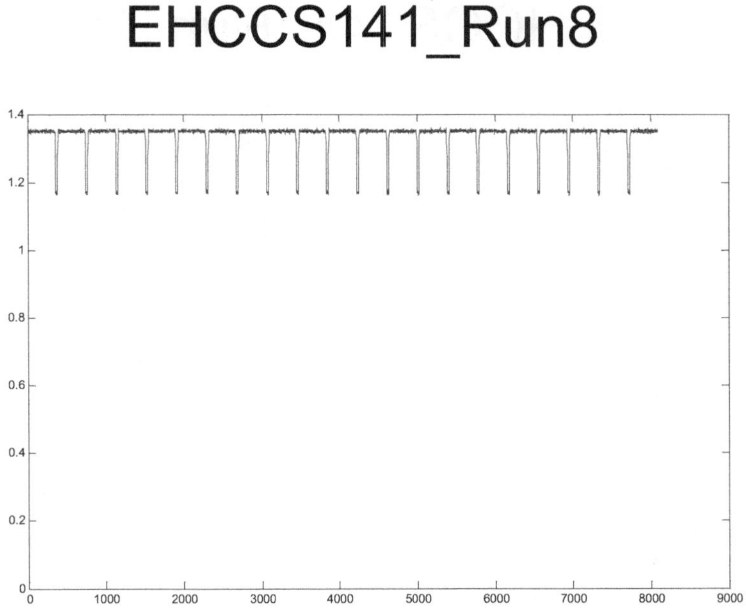

Figure B-23 Baseline Off, 15 min. pulse, Sodium Fluoroacetate

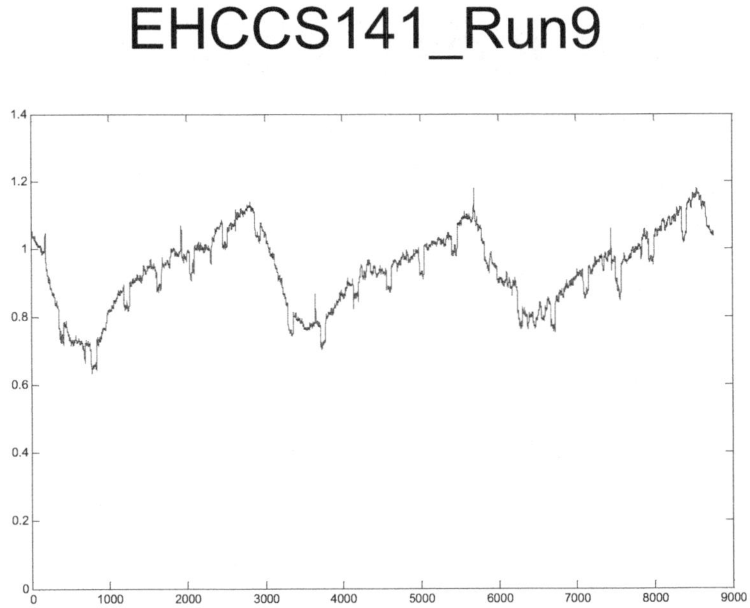

Figure B-24 Baseline On, 30 min. pulse, Low Aldicarb

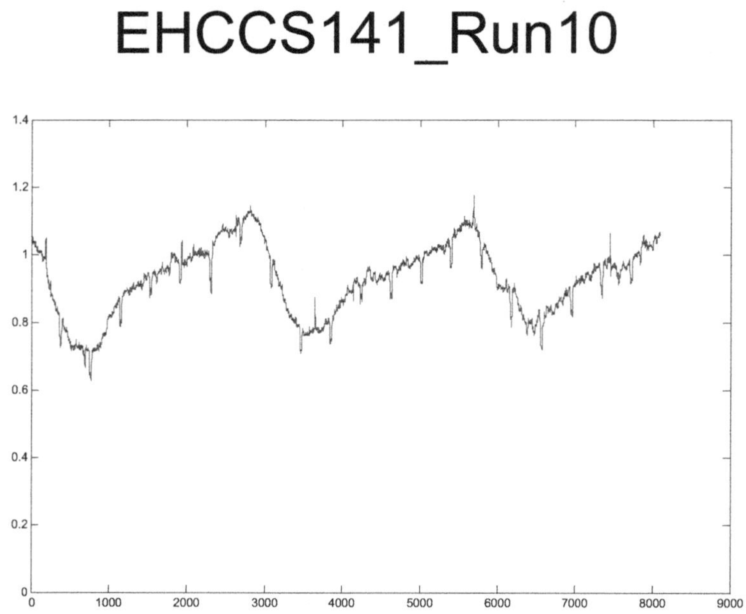

Figure B-25 Baseline On, 15 min. pulse, Low Aldicarb

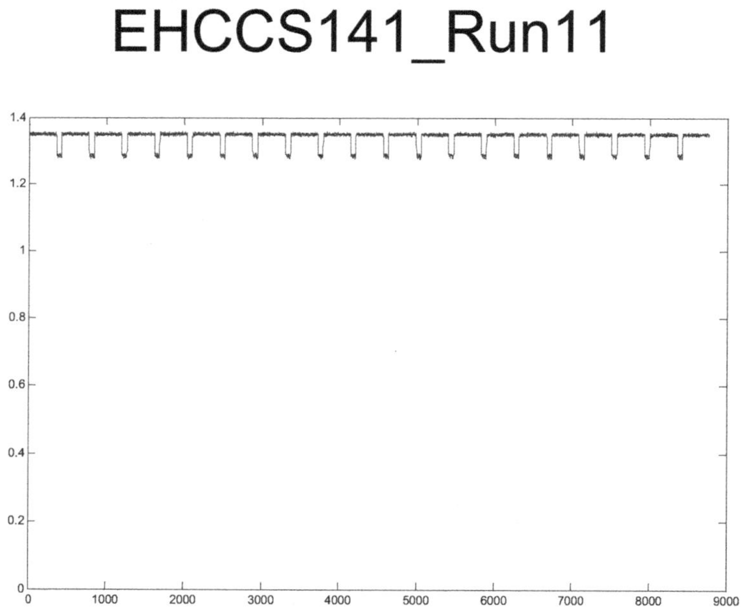

Figure B-26 Baseline Off, 30 min. pulse, Low Aldicarb

EHCCS141_Run12

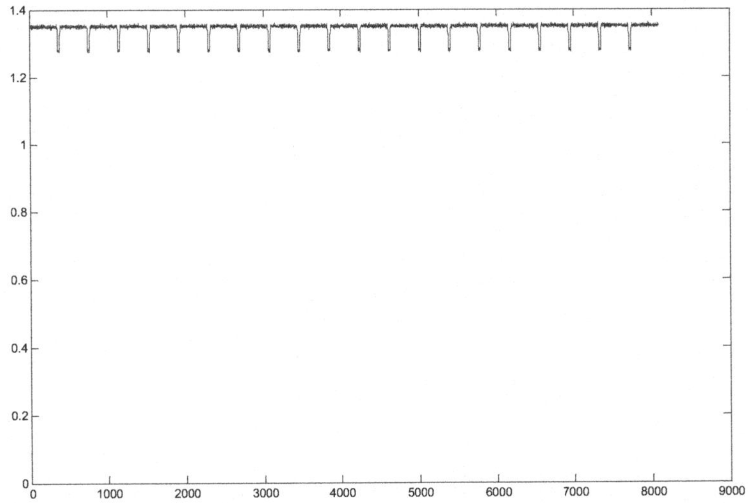

Figure B-27 Baseline Off, 15 min. pulse, Low Aldicarb

EHCCS141_Run13

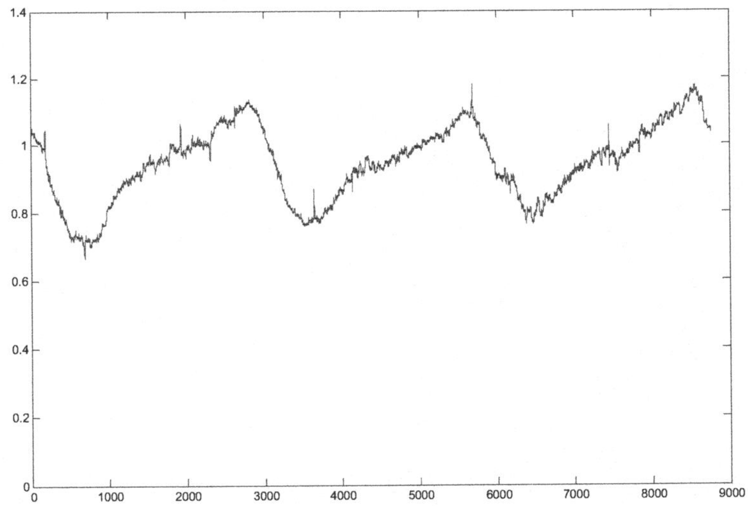

Figure B-28 Baseline On, No Contaminant

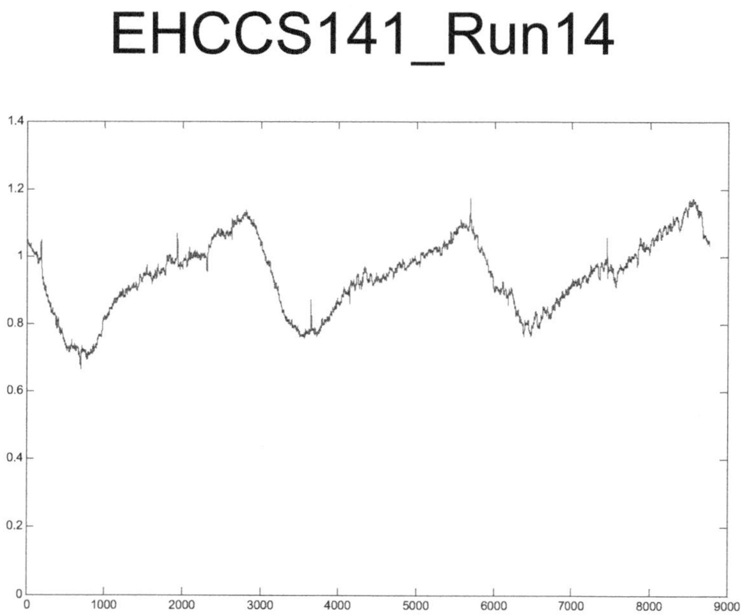

Figure B-29 Baseline On, 30 min. pulse, Low2 Aldicarb

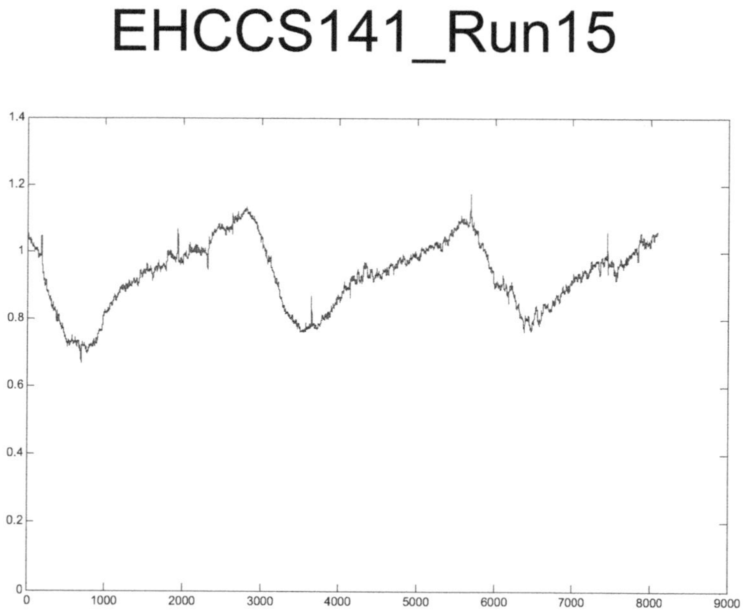

Figure B-30 Baseline On, 15 min. pulse, Low2 Aldicarb

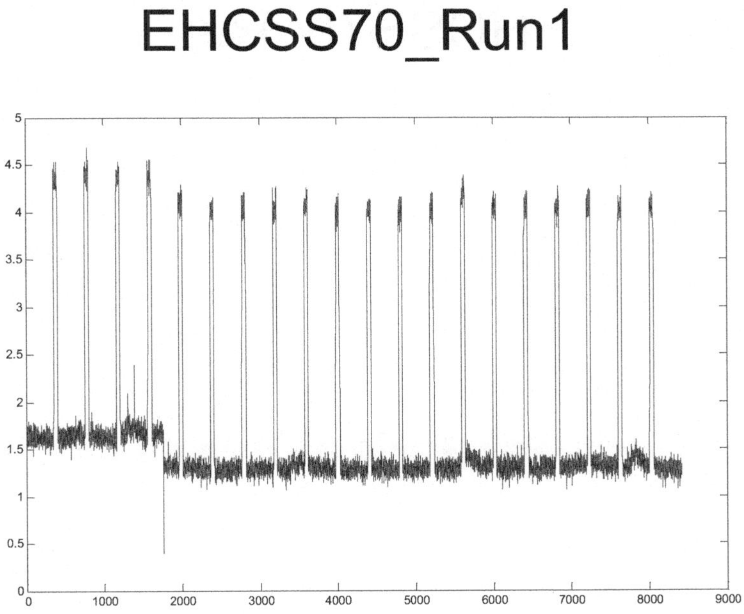

Figure B-31 Baseline On, 30 min. pulse, Aldicarb

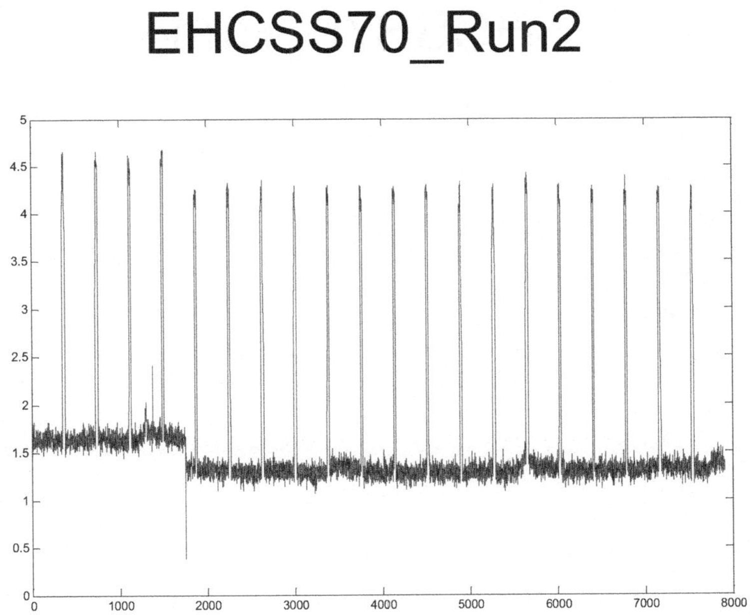

Figure B-32 Baseline On, 15 min. pulse, Aldicarb

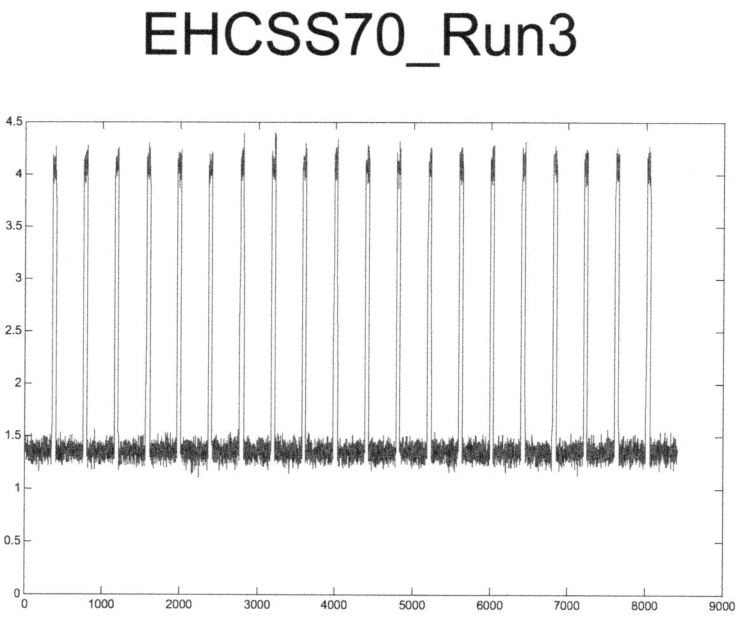

Figure B-33 Baseline Off, 30 min. pulse, Aldicarb

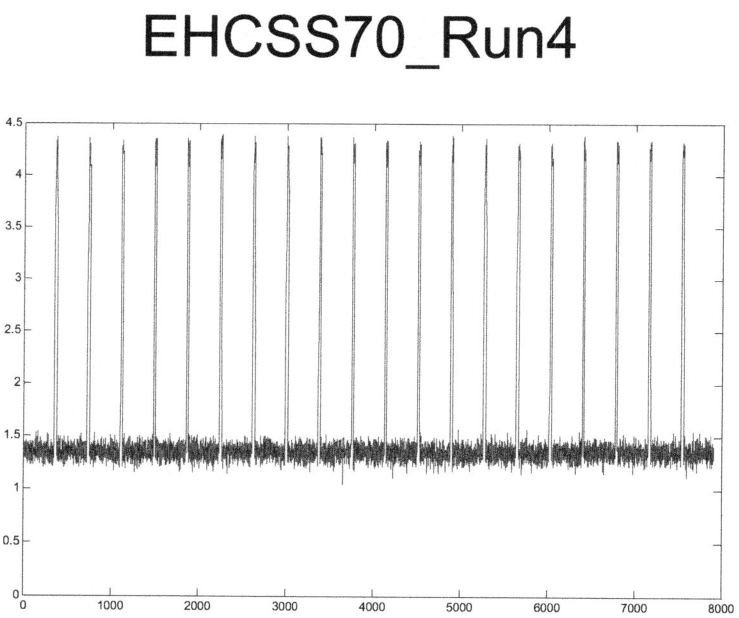

Figure B-34 Baseline Off, 15 min. pulse, Aldicarb

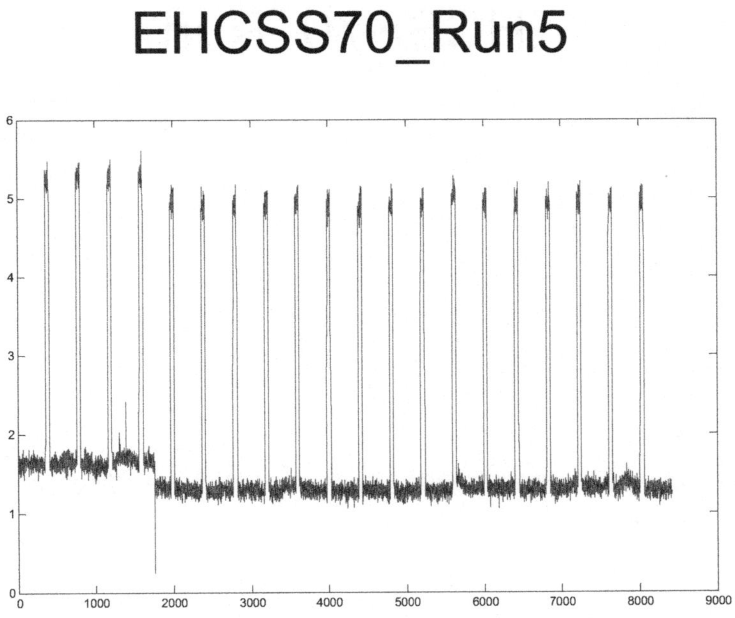

Figure B-35 Baseline On, 30 min. pulse, Sodium Fluoroacetate

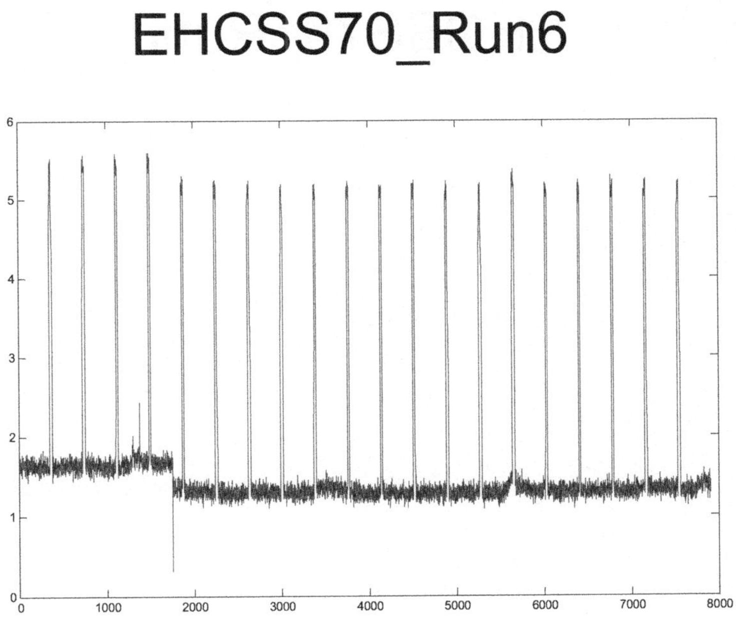

Figure B-36 Baseline On, 15 min. pulse, Sodium Fluoroacetate

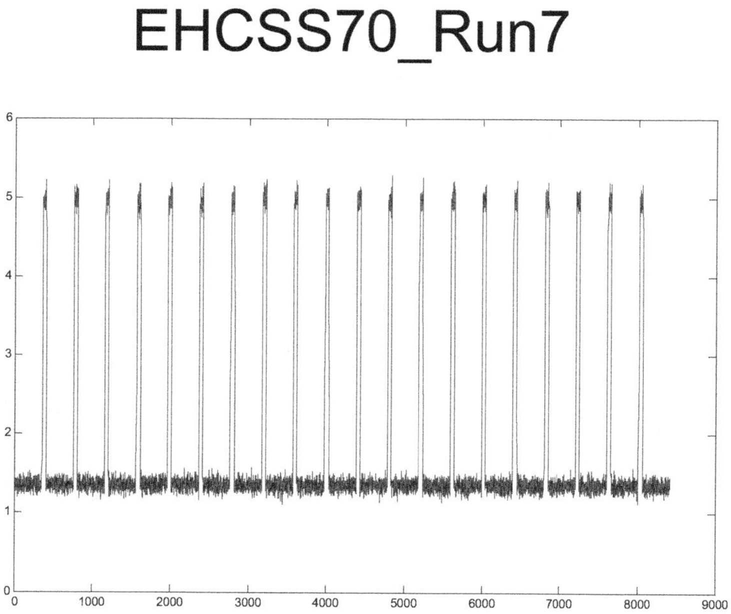

Figure B-37 Baseline Off, 30 min. pulse, Sodium Fluoroacetate

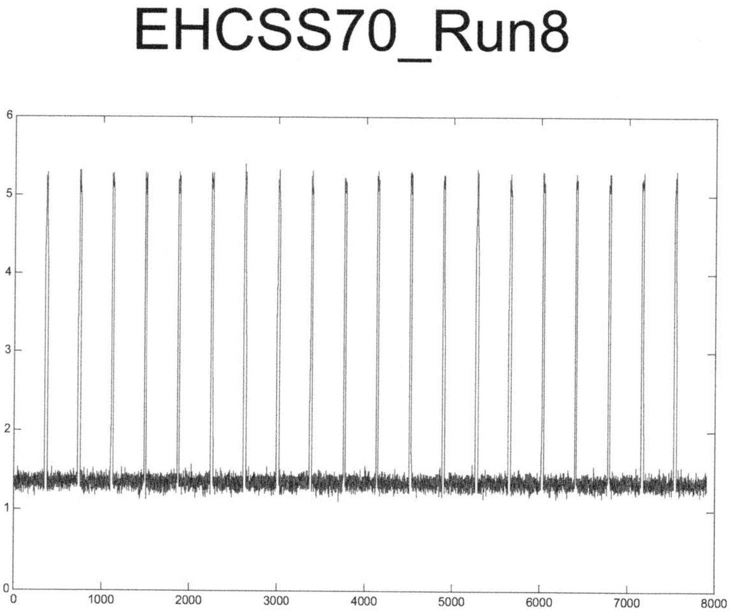

Figure B-38 Baseline Off, 15 min. pulse, Sodium Fluoroacetate

EHCSS70_Run9

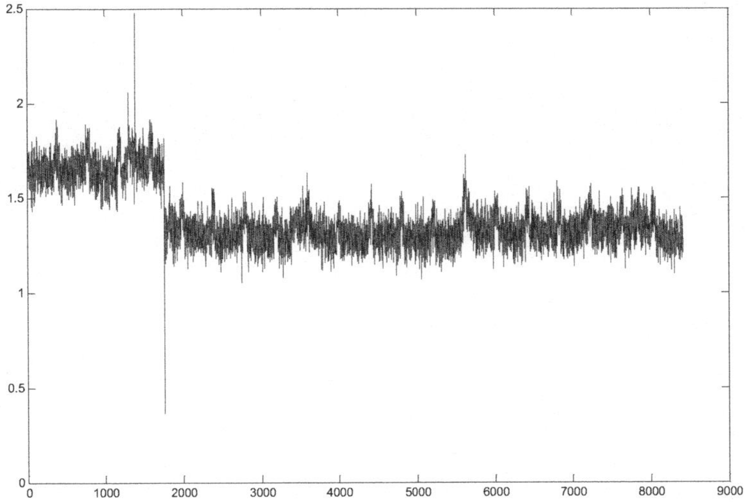

Figure B-39 Baseline On, 30 min. pulse, Low Aldicarb

EHCSS70_Run10

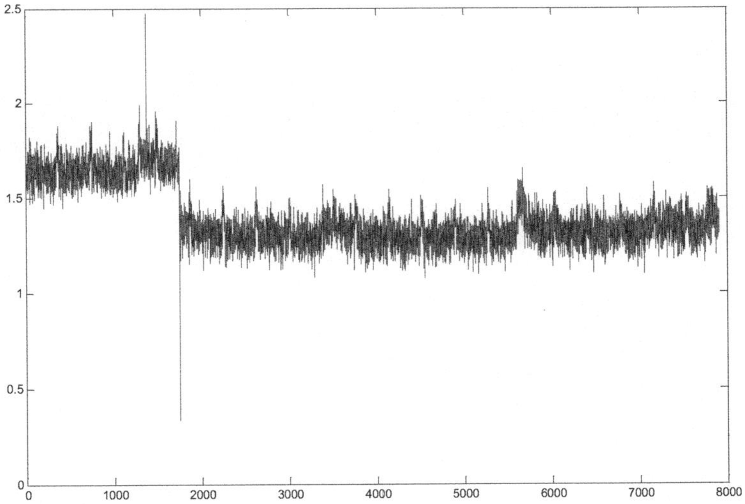

Figure B-40 Baseline On, 15 min. pulse, Low Aldicarb

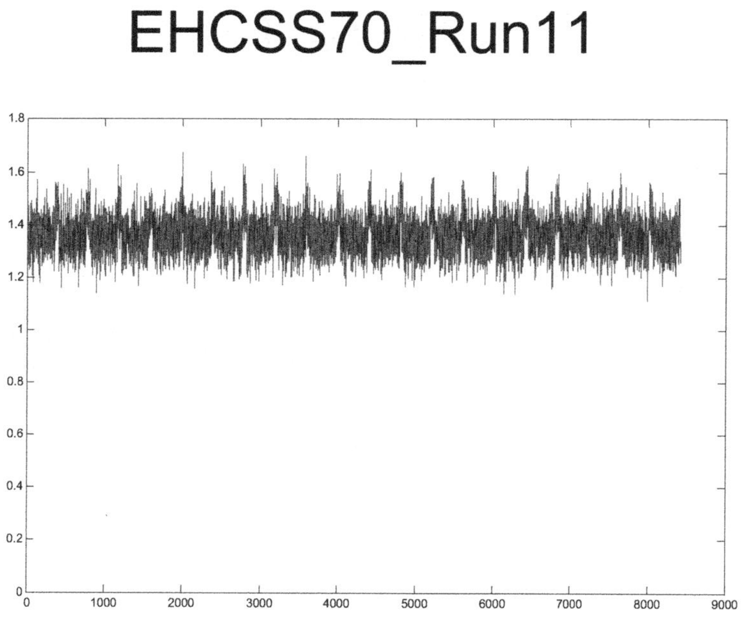

Figure B-41 Baseline Off, 30 min. pulse, Low Aldicarb

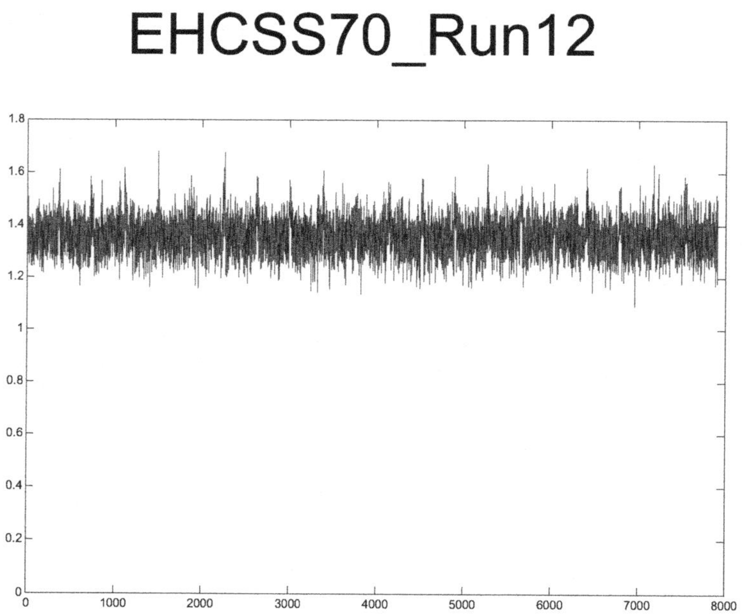

Figure B-42 Baseline Off, 15 min. pulse, Low Aldicarb

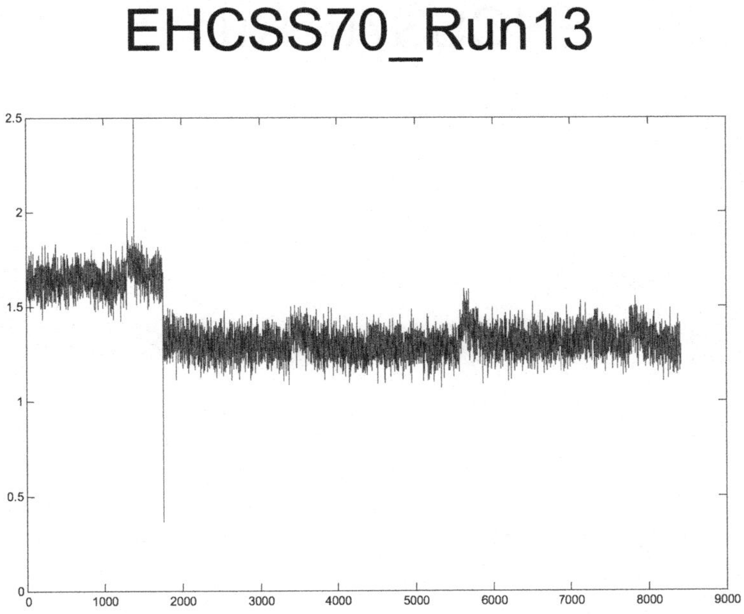

Figure B-43 Baseline On, No Contaminant

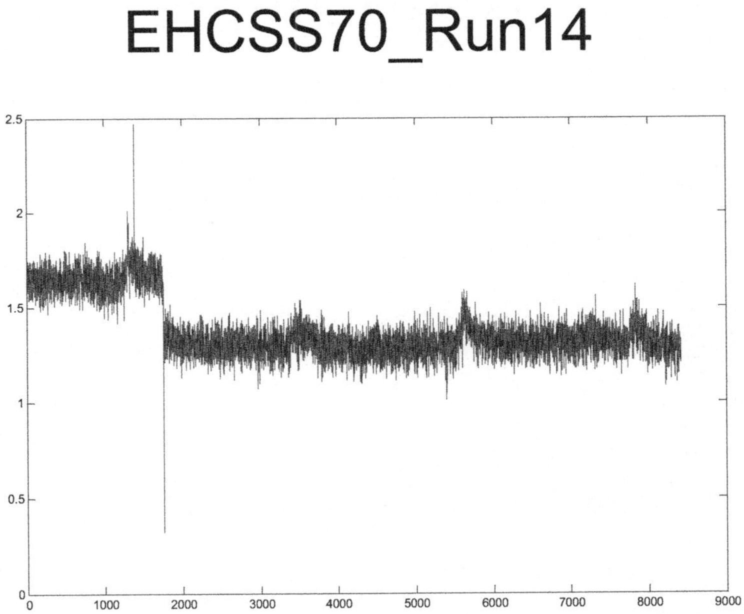

Figure B-44 Baseline On, 30 min. pulse, Low2 Aldicarb

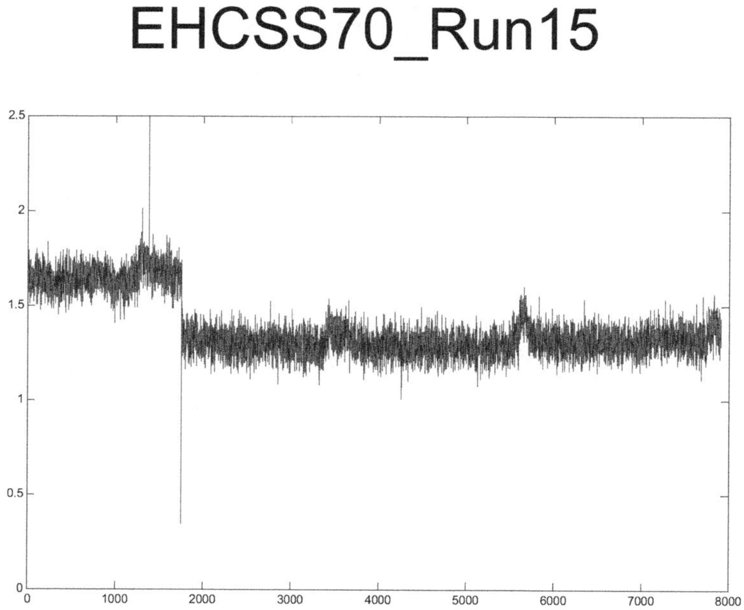

Figure B-45 Baseline On, 15 min. pulse, Low2 Aldicarb

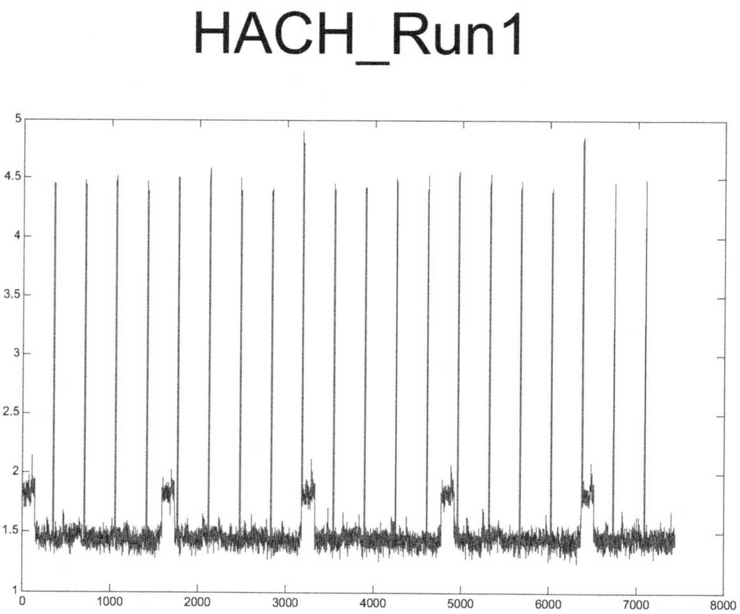

Figure B-46 Baseline On, 30 min. pulse, Aldicarb

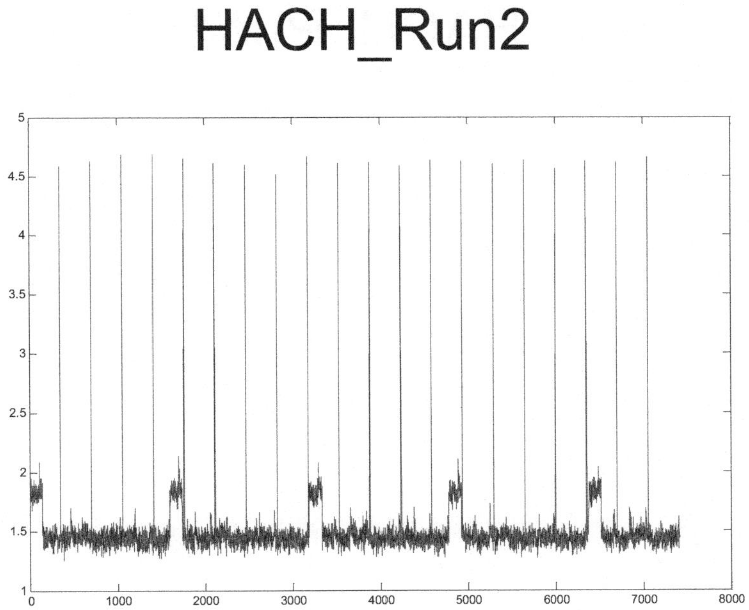

Figure B-47 Baseline On, 15 min. pulse, Aldicarb

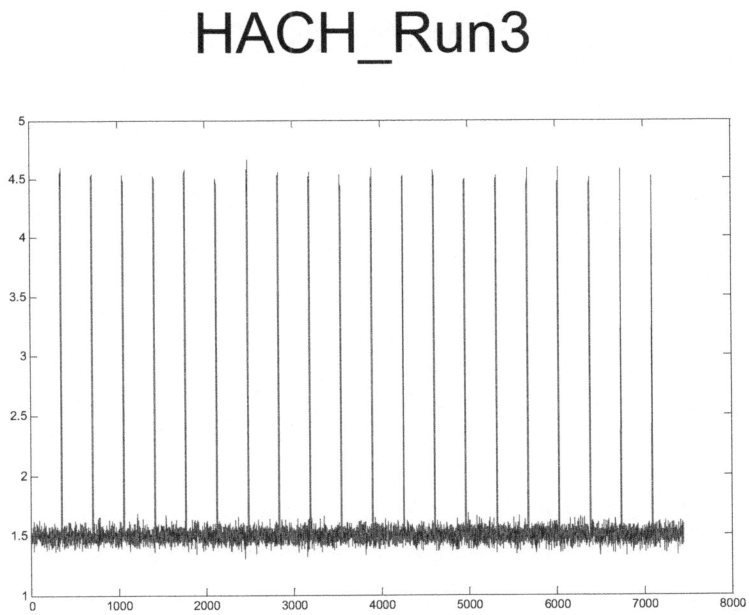

Figure B-48 Baseline Off, 30 min. pulse, Aldicarb

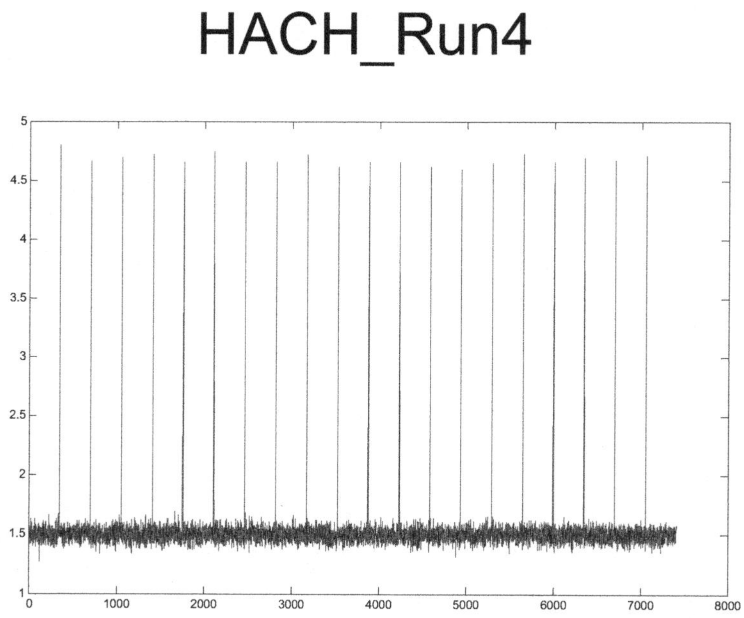

Figure B-49 Baseline Off, 15 min. pulse, Aldicarb

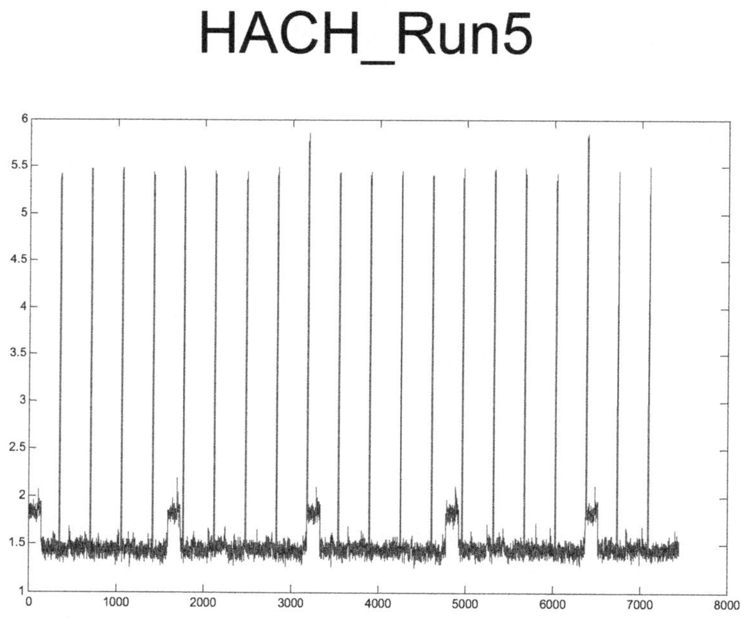

Figure B-50 Baseline On, 30 min. pulse, Sodium Fluoroacetate

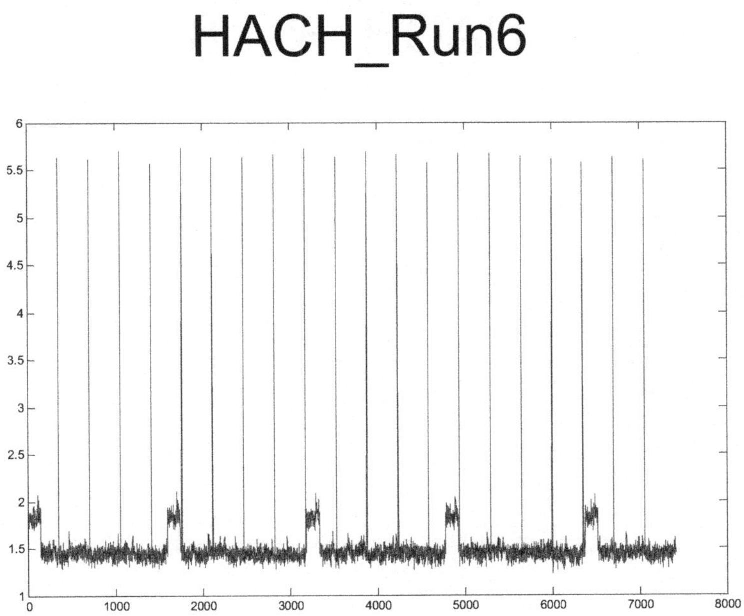

Figure B-51 Baseline On, 15 min. pulse, Sodium Fluoroacetate

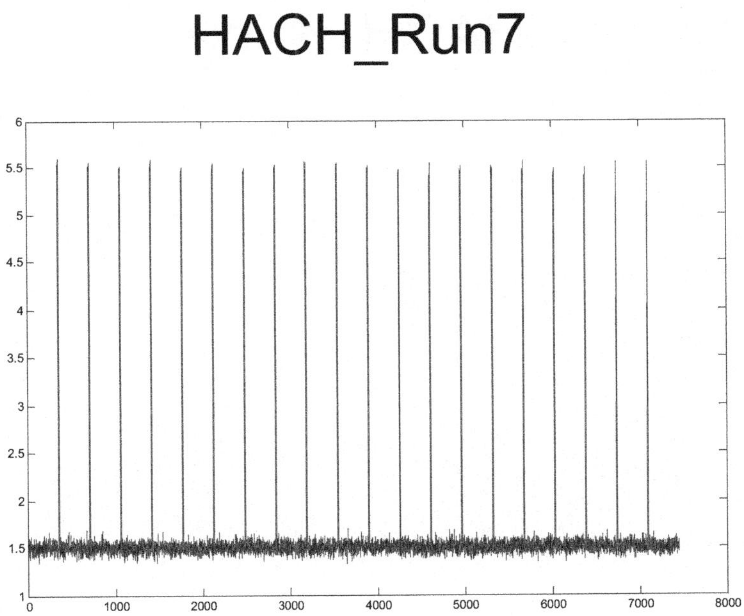

Figure B-52 Baseline Off, 30 min. pulse, Sodium Fluoroacetate

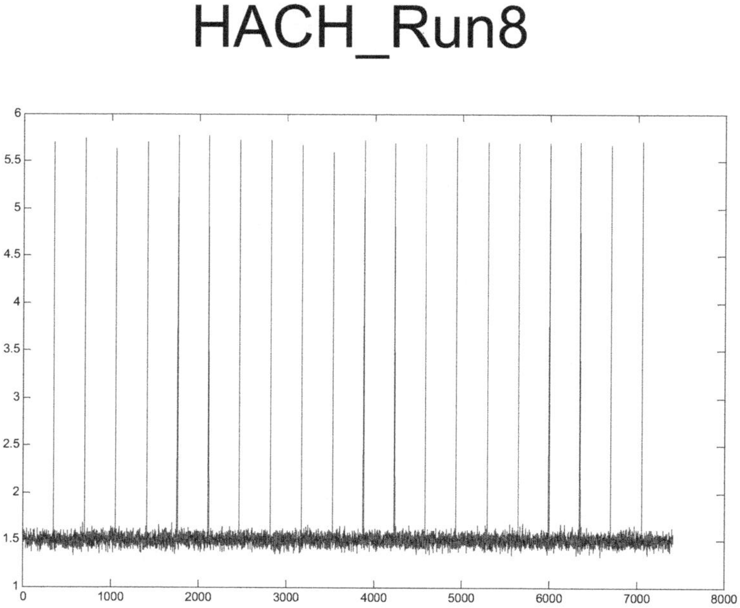

Figure B-53 Baseline Off, 15 min. pulse, Sodium Fluoroacetate

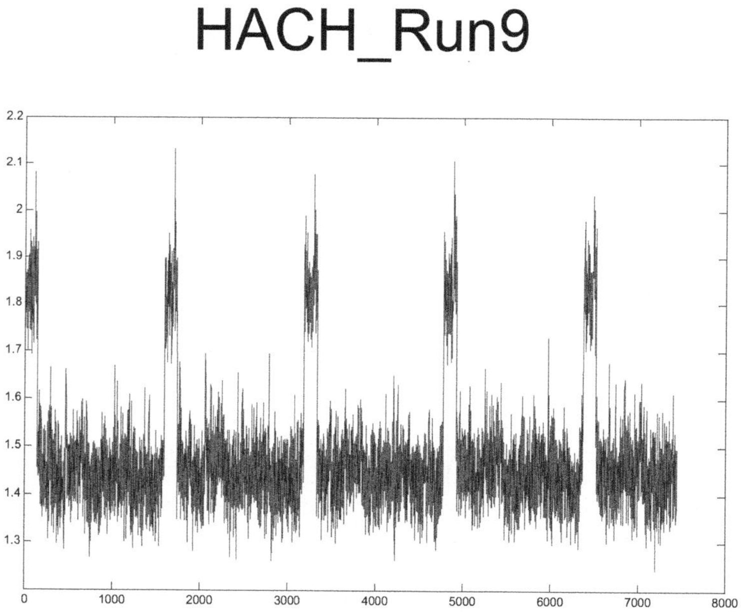

Figure B-54 Baseline On, 30 min. pulse, Low Aldicarb

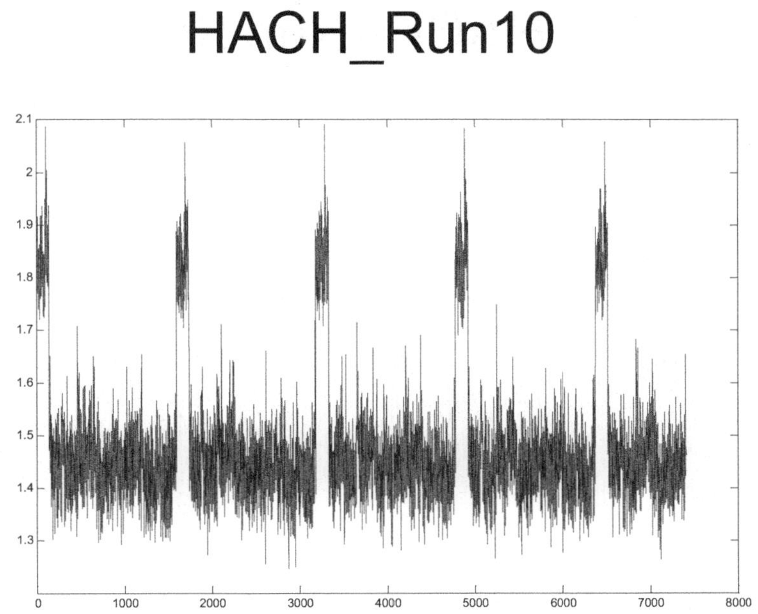

Figure B-55 Baseline On, 15 min. pulse, Low Aldicarb

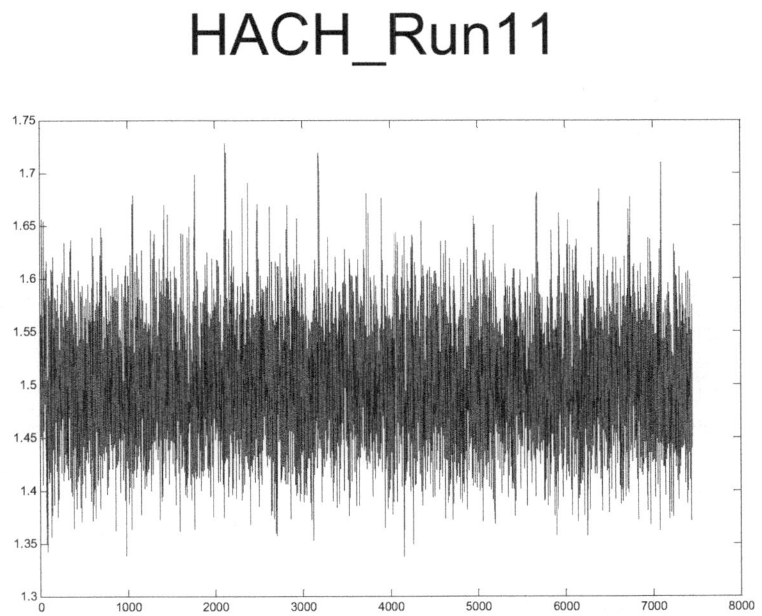

Figure B-56 Baseline Off, 30 min. pulse, Low Aldicarb

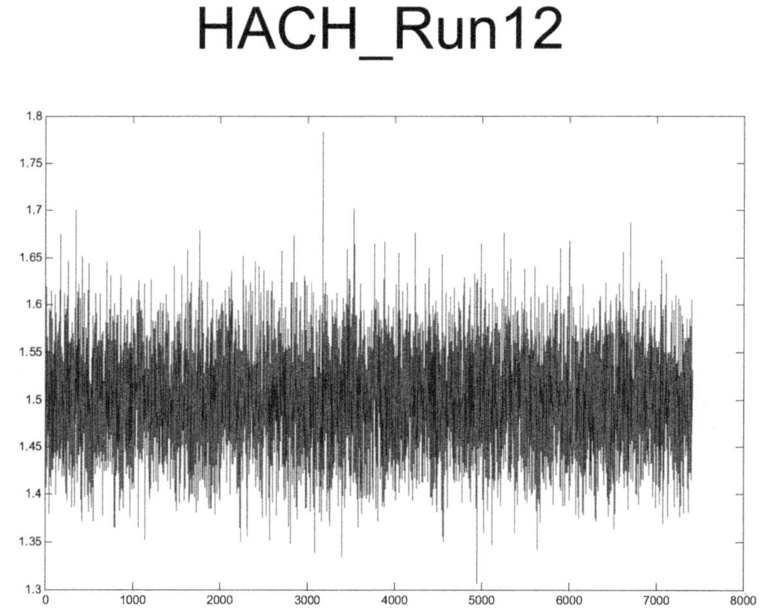

Figure B-57 Baseline Off, 15 min. pulse, Low Aldicarb

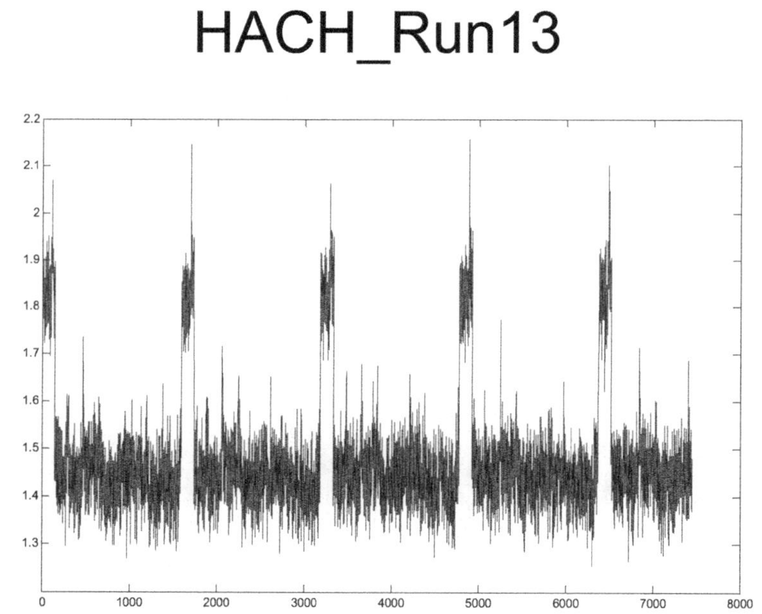

Figure B-58 Baseline On, No Contaminant

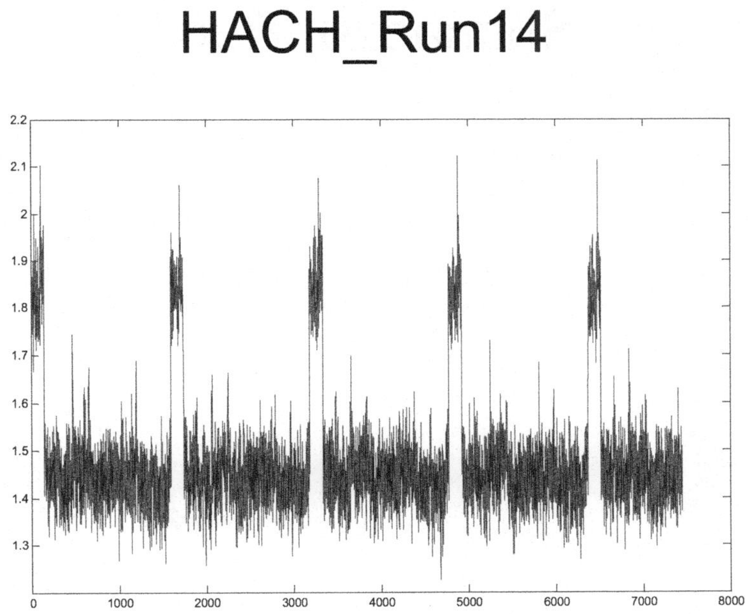

Figure B-59 Baseline On, 30 min. pulse, Low2 Aldicarb

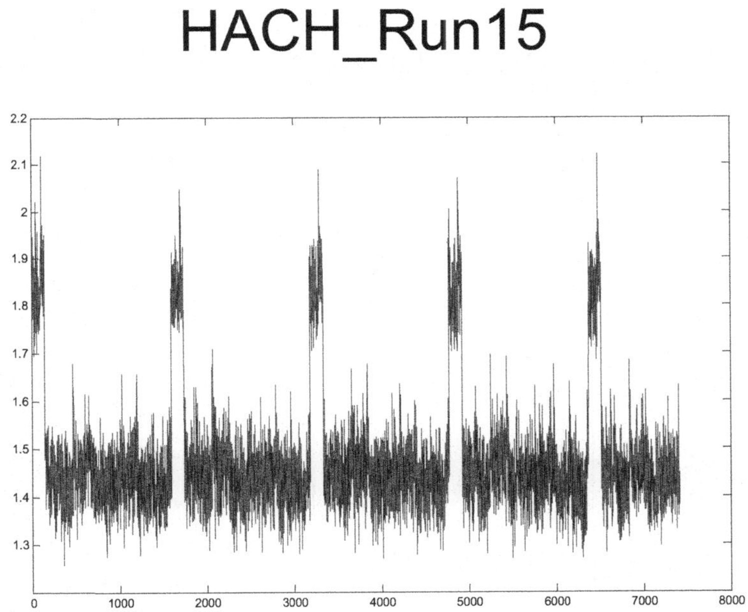

Figure B-60 Baseline On, 15 min. pulse, Low2 Aldicarb

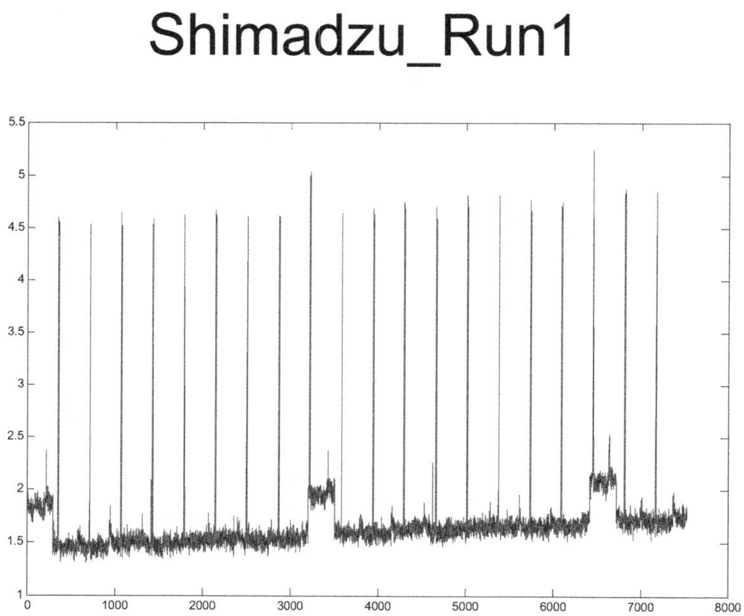

Figure B-61 Baseline On, 30 min. pulse, Aldicarb

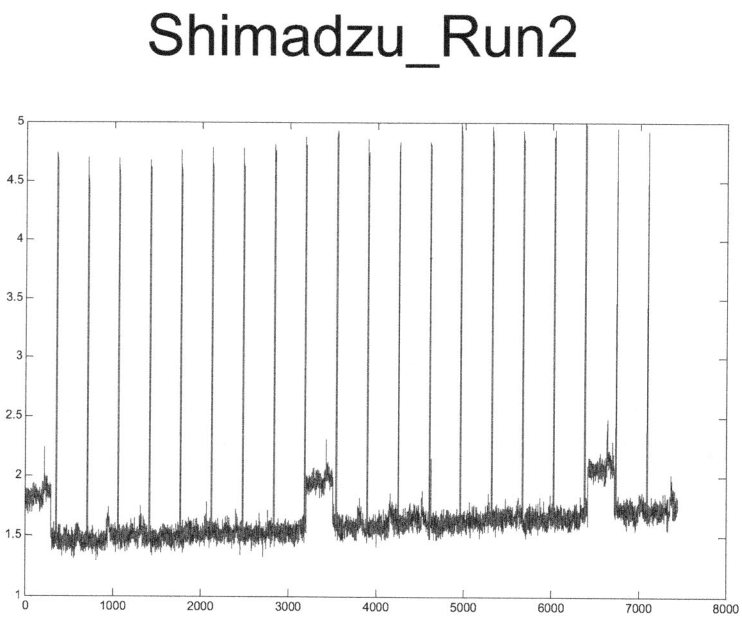

Figure B-62 Baseline On, 15 min. pulse, Aldicarb

Shimadzu_Run3

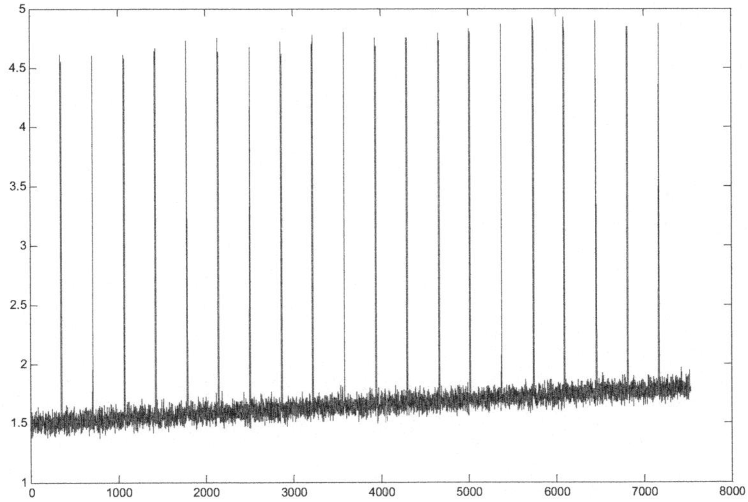

Figure B-63 Baseline Off, 30 min. pulse, Aldicarb

Shimadzu_Run4

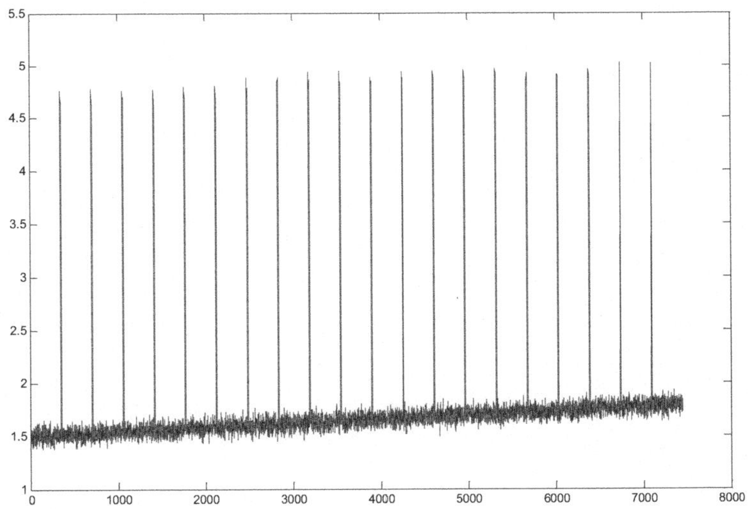

Figure B-64 Baseline Off, 15 min. pulse, Aldicarb

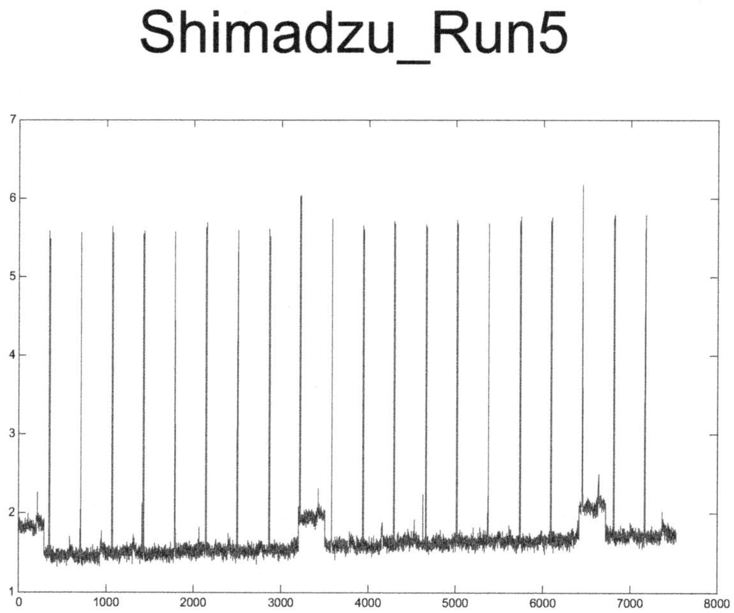

Figure B-65 Baseline On, 30 min. pulse, Sodium Fluoroacetate

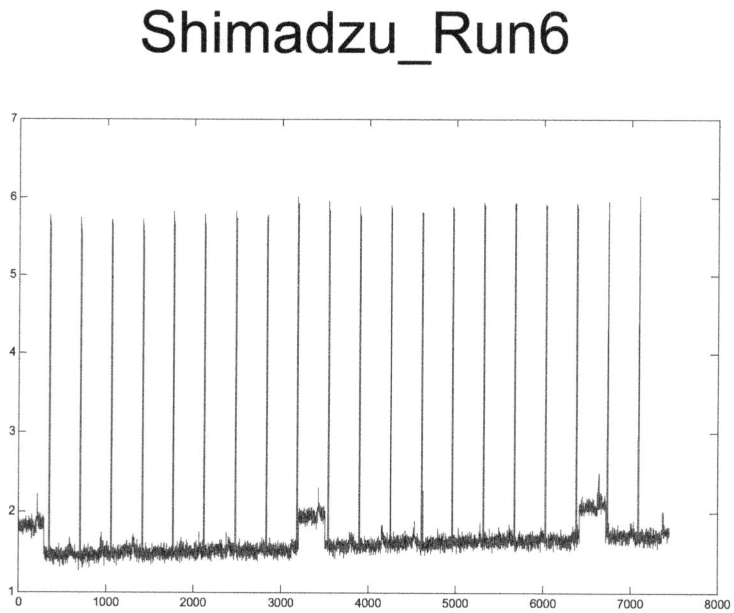

Figure B-66 Baseline On, 15 min. pulse, Sodium Fluoroacetate

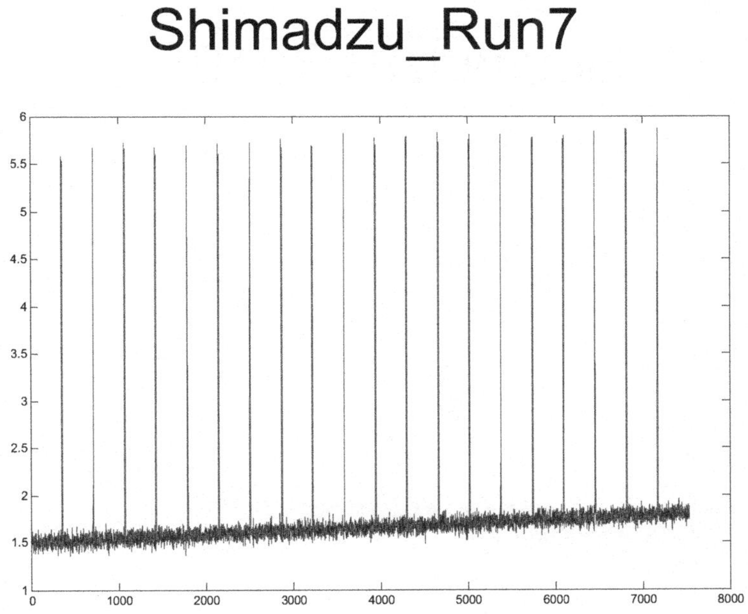

Figure B-67 Baseline Off, 30 min. pulse, Sodium Fluoroacetate

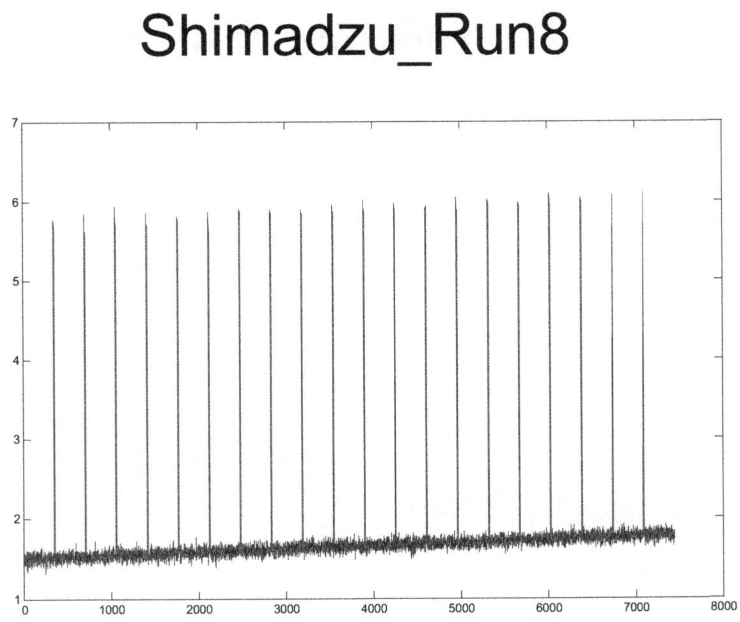

Figure B-68 Baseline Off, 15 min. pulse, Sodium Fluoroacetate

Shimadzu_Run9

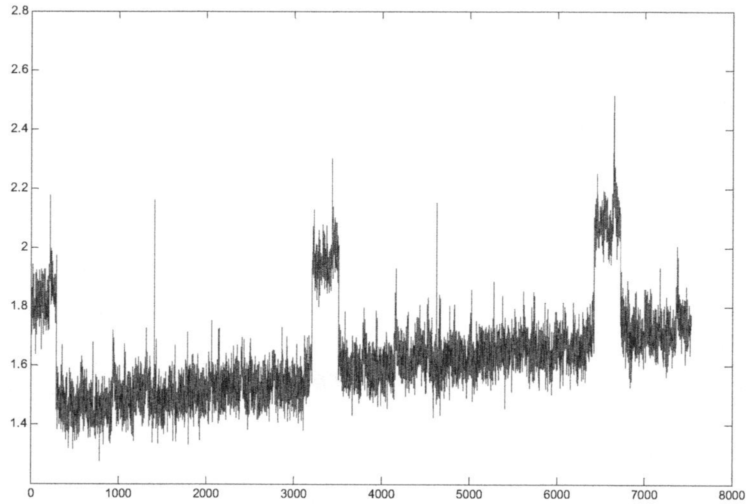

Figure B-69 Baseline On, 30 min. pulse, Low Aldicarb

Shimadzu_Run10

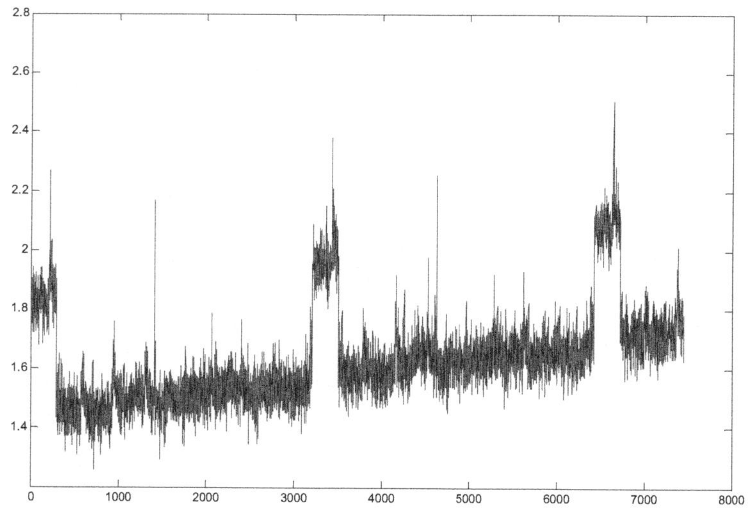

Figure B-70 Baseline On, 15 min. pulse, Low Aldicarb

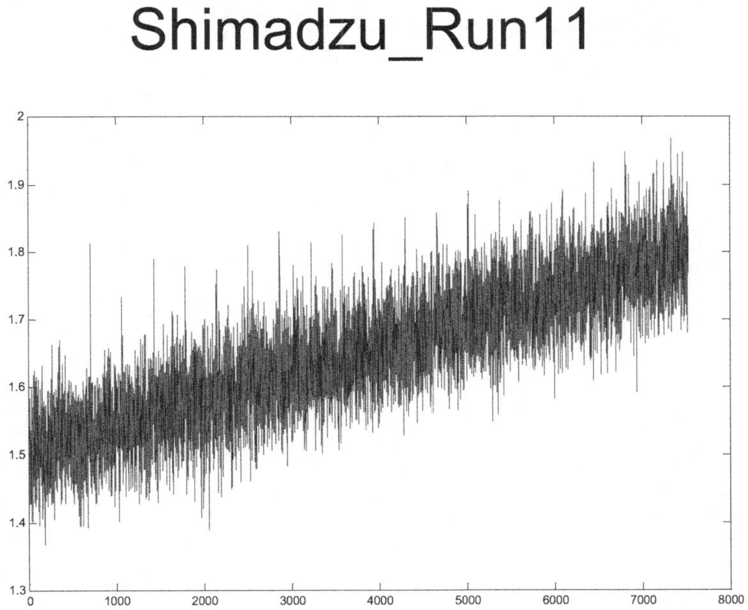

Figure B-71 Baseline Off, 30 min. pulse, Low Aldicarb

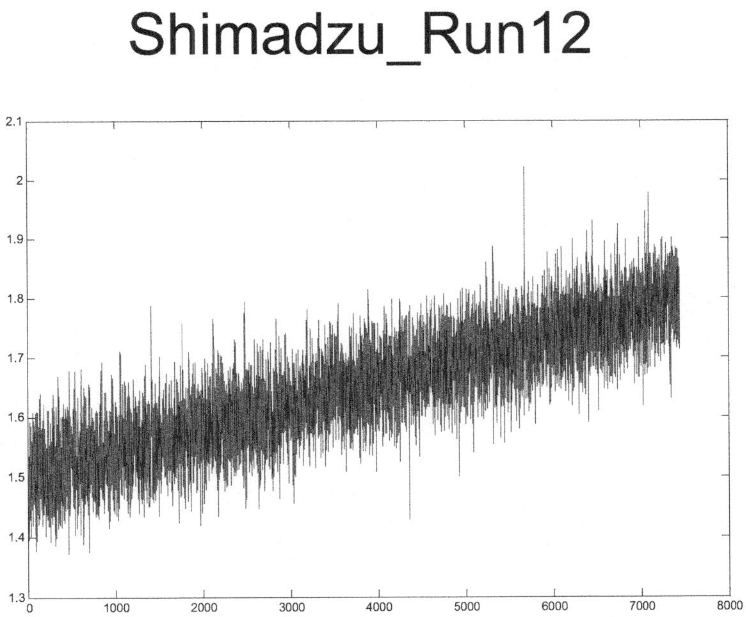

Figure B-72 Baseline Off, 15 min. pulse, Low Aldicarb

Shimadzu_Run13

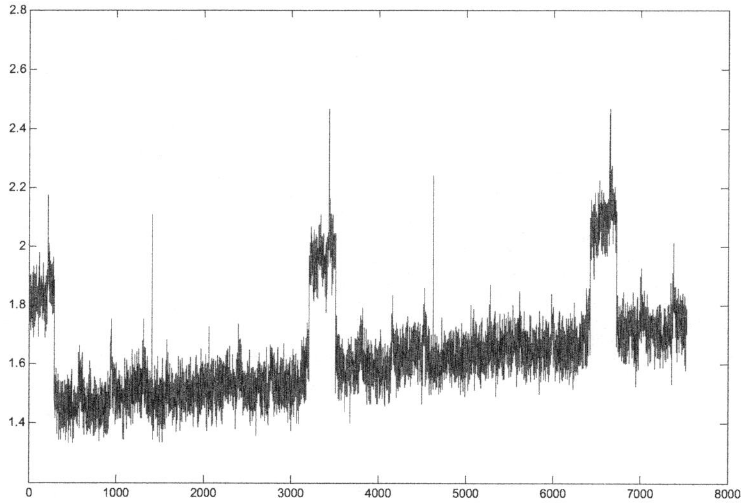

Figure B-73 Baseline On, No Contaminant

Shimadzu_Run14

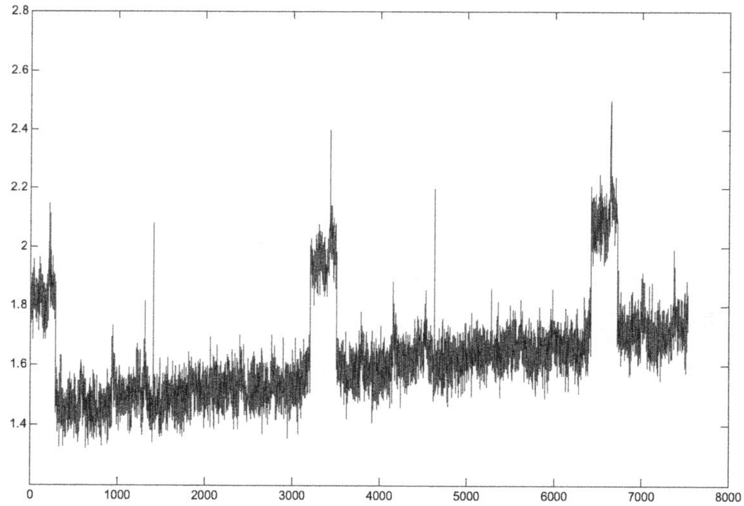

Figure B-74 Baseline On, 30 min. pulse, Low2 Aldicarb

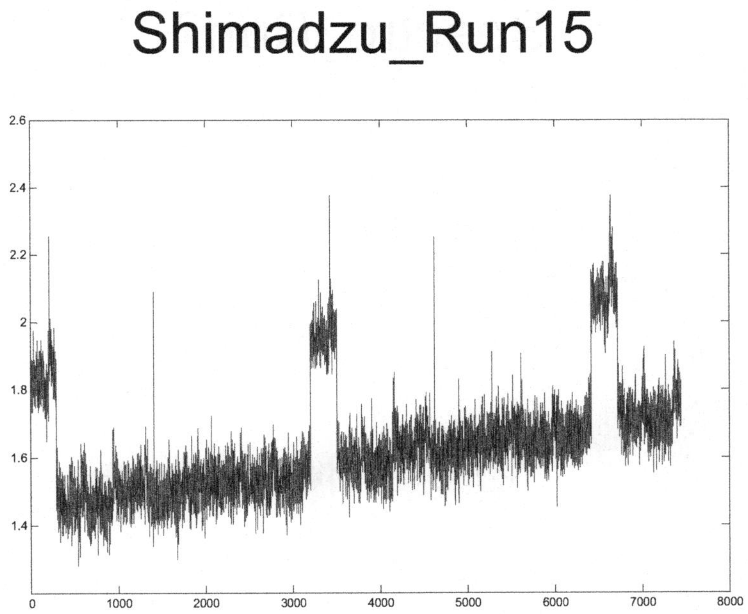

Figure B-75 Baseline On, 15 min. pulse, Low2 Aldicarb

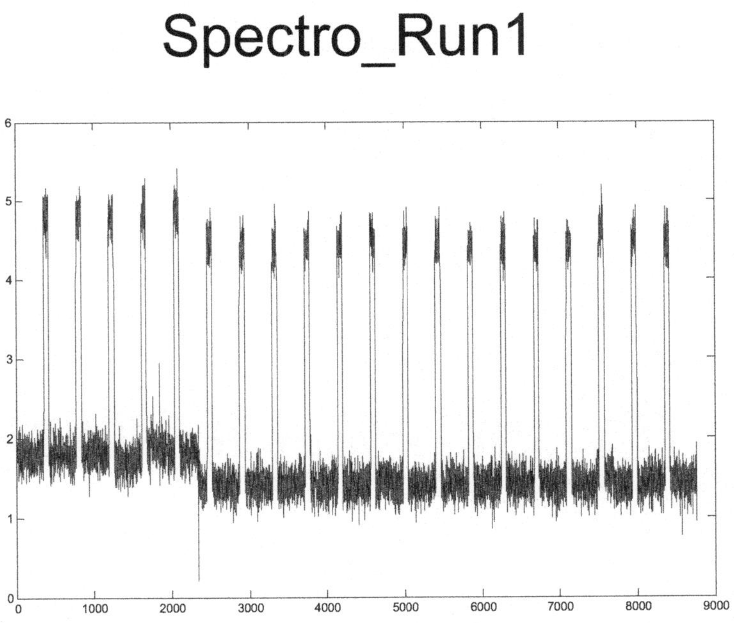

Figure B-76 Baseline On, 30 min. pulse, Aldicarb

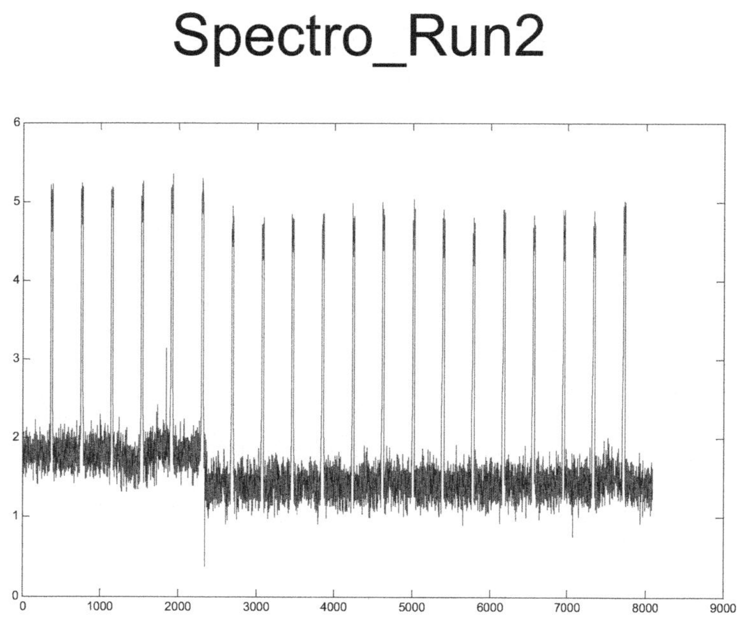

Figure B-77 Baseline On, 15 min. pulse, Aldicarb

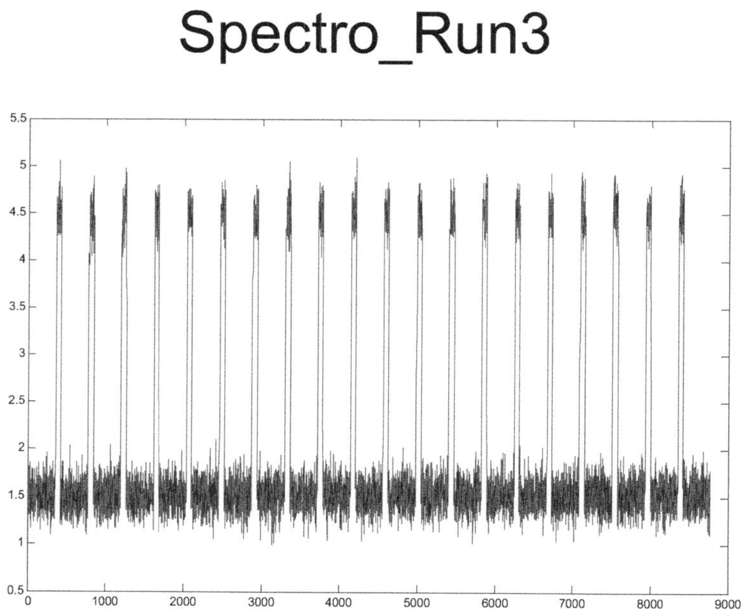

Figure B-78 Baseline Off, 30 min. pulse, Aldicarb

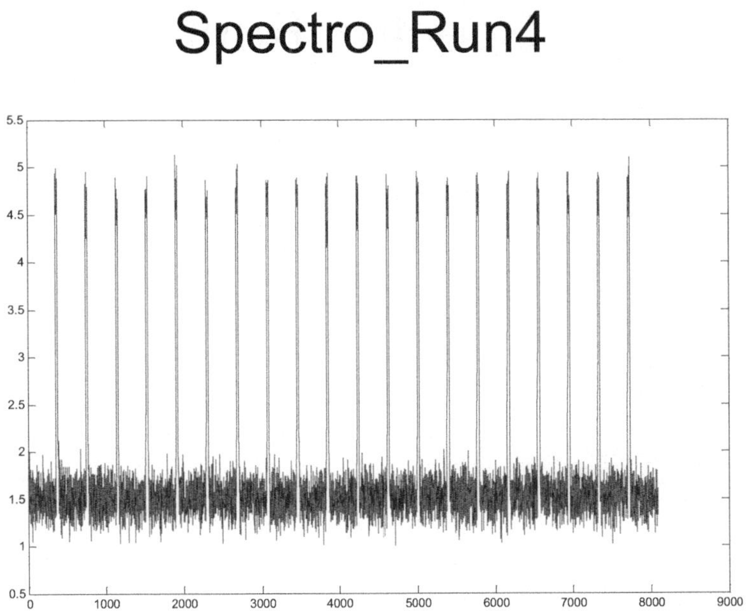

Figure B-79 Baseline Off, 15 min. pulse, Aldicarb

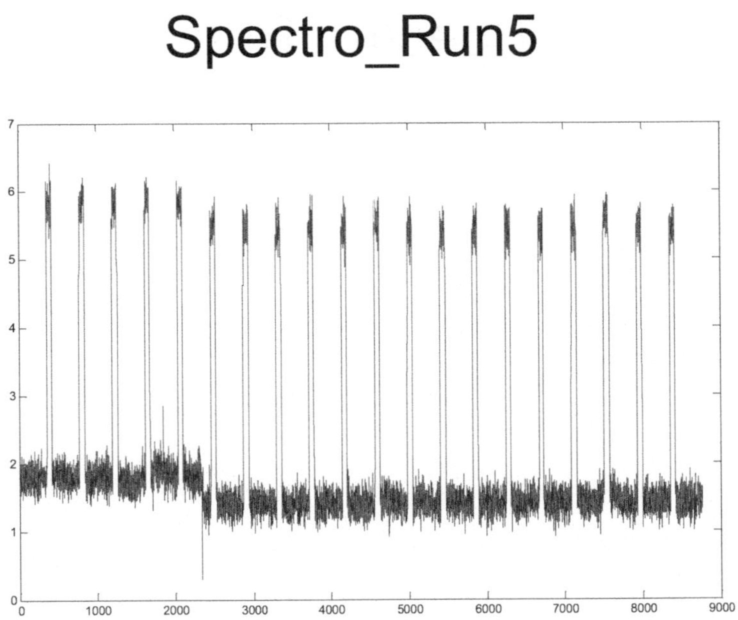

Figure B-80 Baseline On, 30 min. pulse, Sodium Fluoroacetate

Spectro_Run6

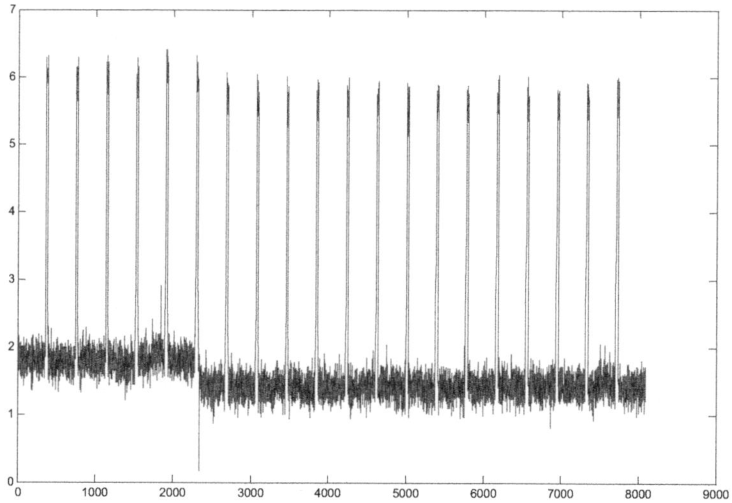

Figure B-81 Baseline On, 15 min. pulse, Sodium Fluoroacetate

Spectro_Run7

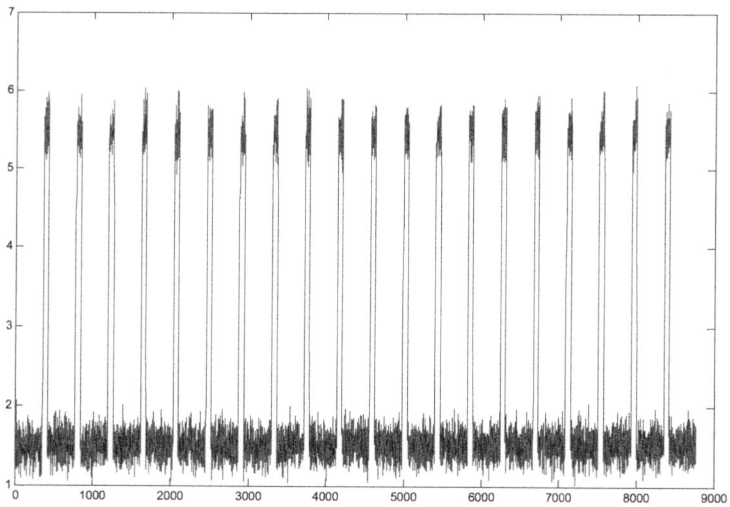

Figure B-82 Baseline Off, 30 min. pulse, Ssodium Fluoroacetate

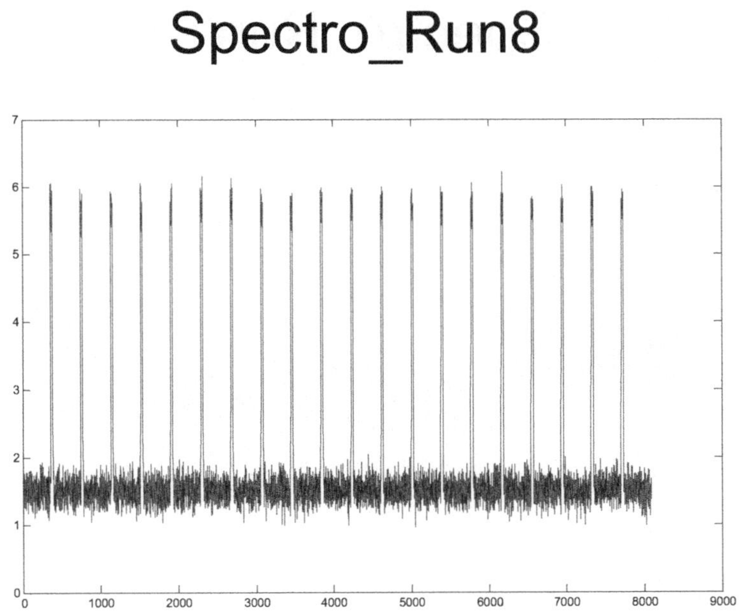

Figure B-83 Baseline Off, 15 min. pulse, Sodium Fluoroacetate

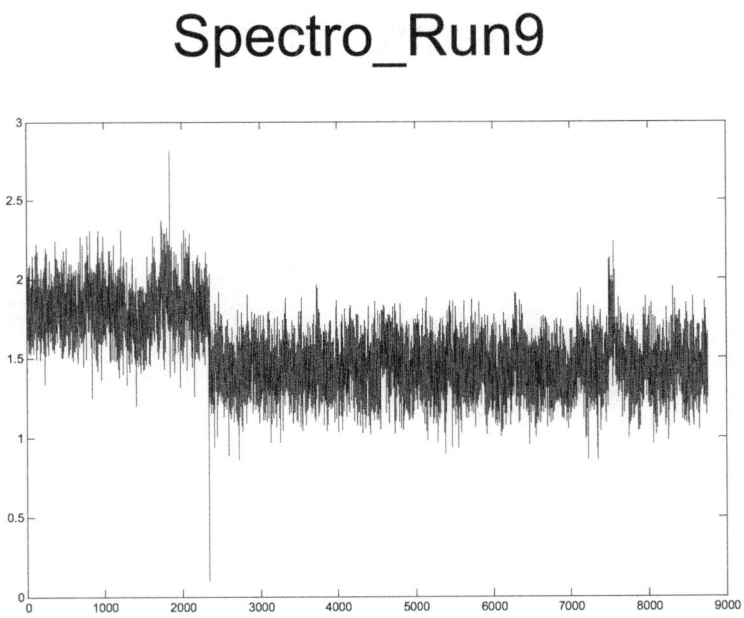

Figure B-84 Baseline On, 30 min. pulse, Low Aldicarb

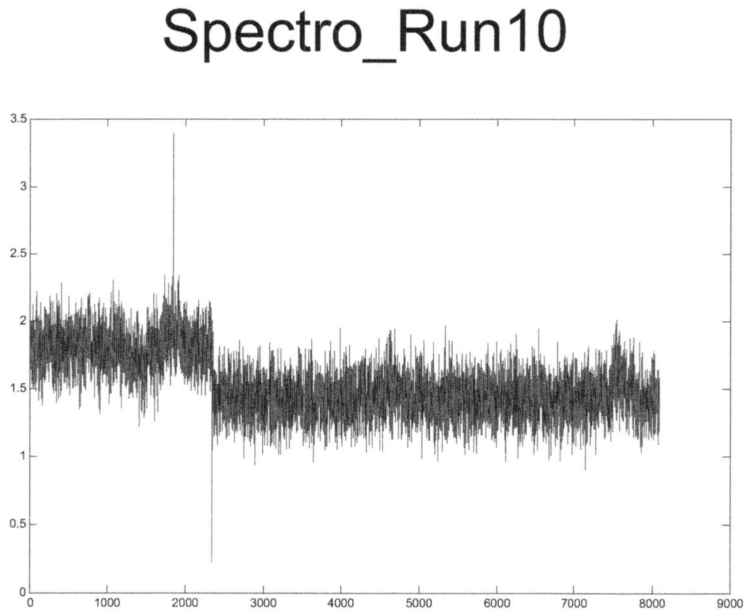

Figure B-85 Baseline On, 15 min. pulse, Low Aldicarb

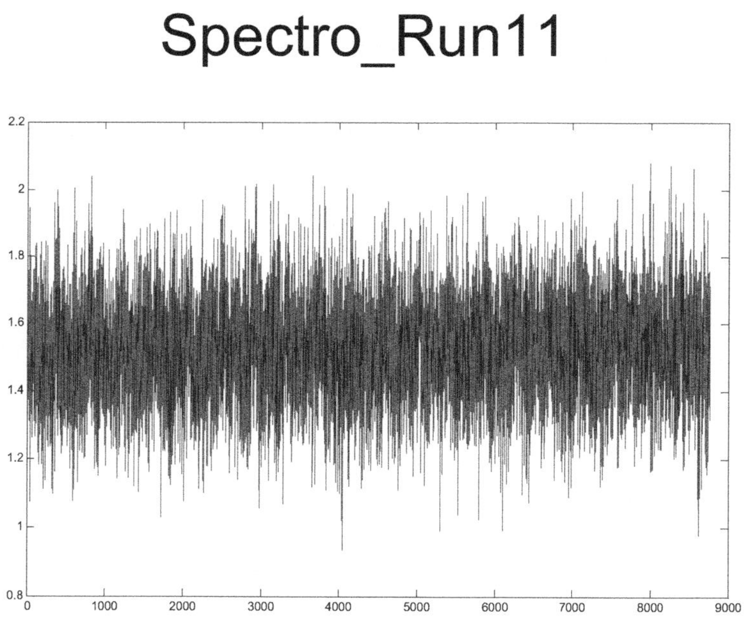

Figure B-86 Baseline Off, 30 min. pulse, Aldicarb

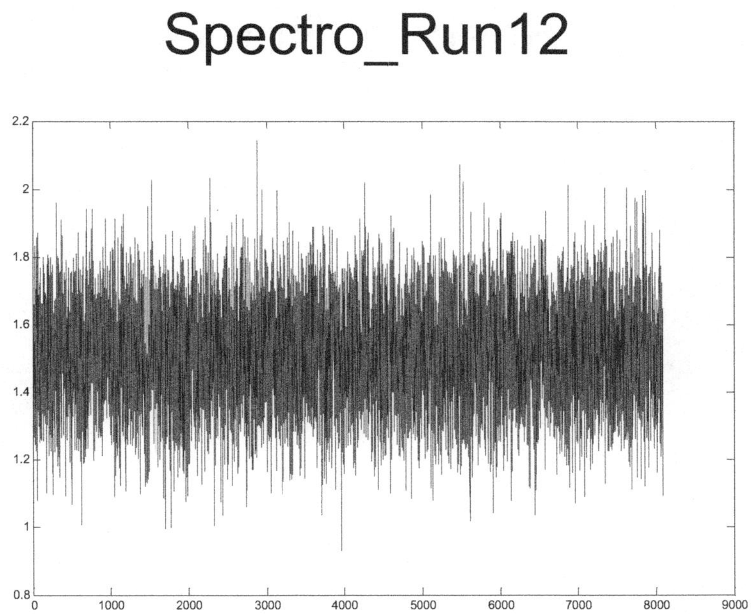

Figure B-87 Baseline Off, 15 min. pulse, Aldicarb

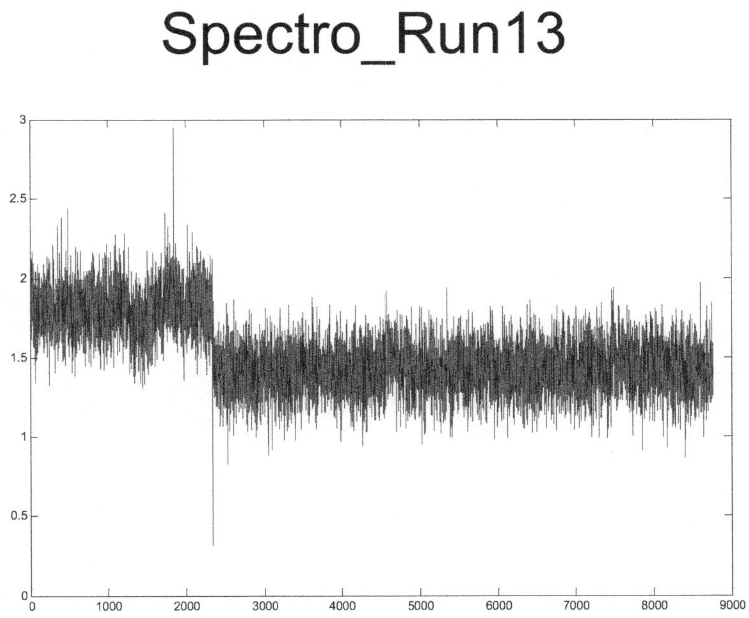

Figure B-88 Baseline On, No contaminant

Spectro_Run14

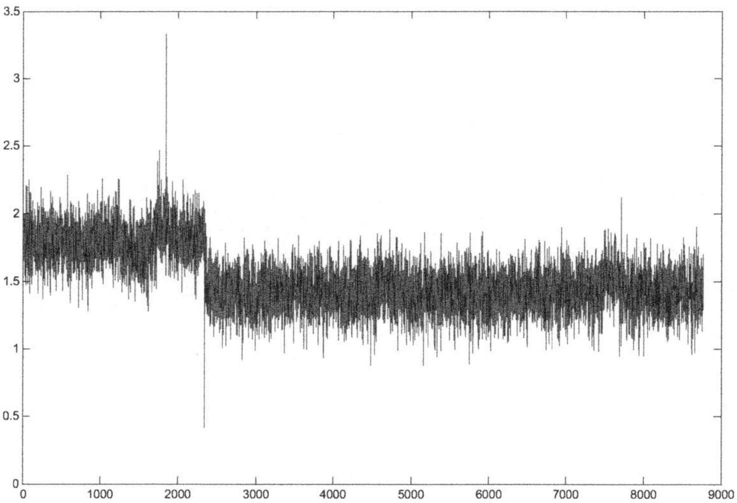

Figure B-89 Baseline On, 30 min. pulse, Low2 Aldicarb

Spectro_Run15

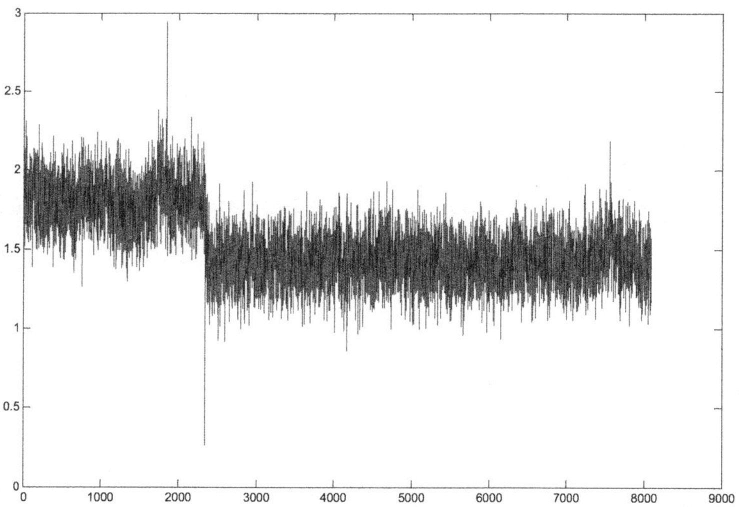

Figure B-90 Baseline On, 15 min. pulse, Low2 Aldicarb

REFERENCES

ATSDR 2007. ATSDR Glossary of Terms. Agency for Toxic Substances & Disease Registry, U.S. Department of Health and Human Services. Available: http://www.atsdr.cdc.gov/glossary.html

Bras, R.L. and I. Rodriguez-Iturbe 1993. *Random Functions and Hydrology*, Dover Publications, Inc., Mineola, New York, 559 pp.

Byer, D., and K. H. Carlson. 2005. Real-time detection of intentional chemical contamination in the distribution system. J. Amer Water Works Assoc. 97(7):130–133.

CBRTA. (Chemical Biological and Radiological Technology Alliance) 2004. Water Monitoring Equipment for Toxic Contamination Technology Assessment. 3M Company, St Paul, MN.

Clark, R. M., N Adam, V. Atluri, M. Halem, E. Vowinkel, P. C. Tao, L. Cummings, and E. Ibrahim. 2004. Developing an Early Warning System for Drinking Water Security and Safety. In *Water Supply Systems Security.* Edited by L. W. Mays. New York, NY: McGraw-Hill Companies.

Conio, O, M. Chioetto, E. Hargeshriner. 2002. Organic Monitors. In *Online Monitoring for Drinking Water Utilities*. Edited by E. Hargeshiriner, Conio and J. Popcovia, Denver, CO: AwwaRF and CRS Proaqua.

EPA 2005a. Technologies and Techniques for Early Warning Systems to Monitoring and Evaluate Drinking Water Quality: A State-of-the-art Review. EPA/600/R-05/156. Available: http://www.epa.gov/NHSRC/pubs/reportEWS120105.pdf

EPA 2005b. Threat Ensemble Vulnerability Assessment Computational Framework, U.S. EPA Homeland Security Research Highlights. Available: http://www.epa.gov/nhsrc/news/news111505b.htm

EPA 2006. Verification Testing Reports: Multi-parameter Water Monitors for Distribution Systems. Available: http://www.epa.gov/etv/verifications/vcenter1-35.html

EPA 2007a. EPANET 2.0. Website. Available: http://www.epa.gov/nrmrl/wswrd/epanet.html

EPA 2007b Technology Testing and Evaluation Program. Available: http://www.epa.gov/nhsrc/tte.htm

EPA 2007c. U.S. EPA Water Security Initiative Web Site. Available: http://yosemite.epa.gov/water/owrccatalog.nsf/9da204a4b4406ef885256ae0007a79c7/cdf35ad0b223d4a7852572d5005f142c!OpenDocument

Foran, J. A. and T. M. Brosnan. 2000. Early Warning Systems for Hazardous Biological Agents in Potable Water. *Environmental Health Perspective*. 108(10): 993-995.

Green, D. M., and J.M. Swets. (1996). Signal Detection Theory and Psychophysics. New York: John Wiley and and Sons Inc.

Gluck, S.J., T. Palmer and M. Ross 2007. Instrumentation and Automation Testing Organizations: How They Work and Why They are Important to Industry. *Journal of Process Analytical Chemistry.* Available: http://www.instrument.org/technical%20articles/JPAC-Testing%20Organizations.pdf

Grayman, W. B., R. A. Deninger, and R. M. Males. 2001. Design of Early Warning and Predictive Source Water Monitoring Systems. Denver, CO: *Awwa Research Foundation*

Hall, J., R.B. Marx, P.C. Cafauber, E. R. Krishnan, R.C. Haught, J.G. Herrmann. 2007. Online Water Quality Parameters as Indicators of Distribution System Contamination. *Jour AWWA* 99:1:66

Hart, D.B., S.A. McKenna, K.A. Klise, V. Cruz and M. Wilson, 2007. CANARY: A Water Quality Event Detection Algorithm Development Tool, in proceedings of: ASCE World Environmental and Water Resources Congress, Tampa, FL, May 2007.

Hart, D.B., S.A. McKenna, K.A. Klise, V. Cruz and M. Wilson, 2007, CANARY: A Water Quality Event Detection Algorithm Development Tool, submitted to: ASCE World Environmental and Water Resources Congress, Tampa, FL, May 15-19th (SAND 2007-3180C)

Haught, R. and M. Fabris. 2002. Inorganic Monitors. In *Online Monitoring for Drinking Water Utilities*. Edited by E. Hargeshiriner, Conio and J. Popcovia, Denver, CO: AwwaRF and CRS Proaqua.

IRIS (Integrated Risk Information System) 2007a. Aldicarb. Available: http://www.epa.gov/IRIS/subst/0003.htm

IRIS (Integrated Risk Information System) 2007b. Sodium Fluoroacetate. Available:IRIS Sodium Fluoro: http://www.epa.gov/iris/subst/0469.htm

ITA (Instrumentation Testing Association) 2006. Available: http://www.instrument.org/

Klise, K.A. and S.A. McKenna, 2006, Water quality change detection: multivariate algorithms, in Proceedings of SPIE (International Society for Optical Engineering), Defense and Security Symposium 2006, April 18-20, Orlando. Florida, 9pp.

Lee. J. Y., and R. A. Deininger. 1999. A Rapid Method for Detecting Bacteria in Drinking Water. *Journal of Rapid Methods and Automation in Microbiology*. 7:135-145.

Lynggaard-Jensen, A. 2002. Specification and Testing of Online Monitors. In *Online Monitoring for Drinking Water Utilities*. Edited by E. Hargeshiriner, Conio and J. Popcovia, Denver, CO: AwwaRF and CRS Proaqua.

McKenna, S.A., K.A. Klise and M.P. Wilson, 2006. Testing Water Quality Change Detection Algorithms, in Proceedings of the 8th Annual Water Distribution System Analysis Symposium, Cincinnati, OH, August 27-30, 2006.

Murray, R., R. Janke, and J. Uber 2004. *The Threat Ensemble Vulnerability Assessment Program for Drinking Water Distribution System Security. Proceedings of EWRI Congress*, Salt Lake City, UT. June, 2004.

Rossman, L., 2000. *EPANET-User's Manual*. United States Environmental Protection Agency (EPA), Cincinnati, OH. Available: http://www.epa.gov/nrmrl/wswrd/dw/epanet.html

Spackman, K. A. (1989). "Signal detection theory: Valuable tools for evaluating inductive learning". *Proceedings of the Sixth International Workshop on Machine Learning*: 160–163, San Mateo, CA: Morgan Kaufman.

States, S., J. Newberry, J. Wichteman, J. Kuchta, M. Scheuring, and L. Casson. 2004. Rapid Analyltical Techniques for Drinking Water Security Investigations. *Journal AWWA* 96:152-164.

Swets, J.A., 1988, Measuring the accuracy of diagnostic systems, *Science*, vol.240, 1285-1293.

ABBREVIATIONS

AUC	Area under (ROC) curve
CWS	Contaminant Warning System
EPA	U.S. Environmental Protection Agency
FLP	Forward Linear Predictor (change detection algorithm)
FN	False negative
FP	False positive
IRIS	Integrated Risk Information System (US EPA)
LOAEL	Lowest observed adverse effect level
mg/L	milligrams per liter
MV	Multivariate (change detection algorithm)
NOAEL	No observed adverse effect level
ppm	Parts per million
ROC	Receiver Operating Characteristic
TEVA	Threat Ensemble Vulnerability Assessment
TN	True negative
TOC	Total organic carbon
TP	True positive
TTEP	Technology Test and Evaluation Program (US EPA)
UV	Ultra-violet

IWA Publishing's authorised EU representative for General Product Safety Regulations is Diane D'Arras, 15 rue Duret, 75116 Paris, France, e-mail: safety@iwap.co.uk.

Printed and bound by CPI Group (UK) Ltd, Croydon, CR0 4YY
17/09/2026
02218802-0015